SURVEYING PRINCIPLES

for Civil Engineers

*Review for the Engineering
Surveying Section of the
California Special Civil
Engineer Examination*

Paul A. Cuomo, PLS

Professional Publications, Inc. Belmont, California

Production Manager: Aline Sullivan Magee
Acquisitions Editor: Gerald R. Galbo
Permissions Editor: Catie Berkenfield
Copy Editor: Mia Laurence
Book Designer: Charles P. Oey
Typesetter: Cathy Schrott
Illustrator: Charles P. Oey
Proofreader: Jessica R. Whitney-Holden
Cover Designer: Charles P. Oey
Cover Photographer: Charles P. Oey, with special thanks to Haselbach Surveying Instruments

SURVEYING PRINCIPLES FOR CIVIL ENGINEERS:
Review for the Engineering Surveying Section of the California Special Civil Engineer Exam

Printed in the United States of America

Professional Publications, Inc.
1250 Fifth Avenue, Belmont, CA 94002
(650) 593-9119
www.ppi2pass.com

Current printing of this edition: 2

Library of Congress Cataloging-in-Publication Data
Cuomo, Paul A.
    Surveying principles for civil engineers: review for the
  engineering surveying section of the California special civil
  engineer examination / Paul A. Cuomo.
        p.   cm.
    ISBN 1-888577-08-8
    1. Civil engineering--California--Examinations--Study guides.
  2. Surveying--California--Problems, exercises, etc.   I. Title.
  TA159.C86   1997
  526.9'076--DC21   97-12592
                                                         CIP

Table of Contents

Preface

This book was created to be used as a study guide for the engineering surveying portion of the California Special Civil Engineer Examination. It has been prepared with that goal in mind; however, this book includes a diversity of subjects that are relevant to the practices of civil engineers, and will be a valuable reference to almost anyone who practices civil engineering in California.

The need for this book is apparent. Civil engineering students are offered few surveying courses, and consequently, the passing rate for the special engineering surveying portion of the exam is low. This book was created to supplement your surveying education. In addition, you are encouraged to attend basic surveying classes and review courses that focus on problem-solving so you will be prepared to answer the material on the day of the test.

Overall, I feel I have captured the essence of what a modern civil engineer needs to know. This book can be used as both an exam preparation tool and as a reference throughout a civil engineer's career.

Paul A. Cuomo, PLS

Acknowledgments

Needless to say, this book is not the fruit of a single effort on my part. There is a long list of persons and companies who have contributed to this publication. It is difficult to know where to start acknowledging anyone without the appearance of an order of importance to their roles, so I will list these in alphabetical order and hope to stay out of trouble. Much thanks and appreciation to:

- George Dorman, Photogrammetrist, for his advice in putting together and for reviewing the "Photogrammetry" chapter.

- Jeremy Evans, PLS, for preparing the chapters "Error Analysis" and "California Coordinate System" and for modifying and preparing problems for the "Astronomy" chapter. These are outstanding chapters.

- Carol Landmann, PLS, for her extraordinary efforts in preparing the text and many of the illustrations from my sloppy, hard-to-read handwritten notes, and for her dogged attempts to keep my grammar acceptable.

- Roy Minnick, PLS, author and publisher for his daily advice on everything I needed to know to do this book and then some.

- John D. Pavlik, PLS, for preparing the chapter "Vertical Distance." John used his many years of teaching experience to produce a fine chapter.

- Mark Pfieler, PLS, for allowing the use of a portion of his firm's *Handbook of Construction Survey Staking* as part of the "Construction Staking" chapter.

- Ed Zimmerman, PLS, and author, for helping me put together material for the "Traversing" chapter.

- And last but not least, my wife Peggy, for her support and her prodding, pushing, cajoling, and general encouragement to get to work on this book on days when it seemed impossible.

I am grateful for all the help and good work of these contributors.

How To Use This Book

This review manual is based on the state of California's Board of Registration for Professional Engineers and Land Surveyors special surveying examination for civil engineers. Its content is based on the following guidelines as provided by the Board:

A. Engineering Surveying Equipment and Field Activities

B. Engineering Surveying Field Measurements

C. Engineering Surveying Calculations

D. Engineering Surveying Office Procedures

This 2½-hour multiple-choice examination is administered the day after the 8-hour civil examination. The point value for each of the problems is given in the problem statements. Passing scores on the official examination vary. For the examination given in October 1996, a raw score of 176 points out of a possible 287 points (or 61%) was required to pass. For the examination given in April 1997, a raw score of 166 points out of a possible 275 points (or 60%) was required to pass. Historical passing scores for examinations prior to April 1997 should be disregarded since the examination format and grading changed with the April 1997 examination.

This review manual is intended to serve as the foundation for preparation for the special surveying examination, serving to both define scope and provide practice in problem solving.

The technical chapters of this book contain narrative, examples, and problems with solutions. I hope the user will try to solve the problems using the text before looking at the solutions. This is a good practice to follow in preparation for an open-book exam.

Surveying and California Law 1

1. Introduction

A. Types of Surveys

Land surveying is described based on which features of the land are being measured. The types of surveys are degined as follows.

- *Boundary surveys* document property lines.

- *Geodetic surveys* measure the shape of the earth from a planetary-wide perspective.

- *Hydrographic surveys* map the features and shapes located under a body of water.

- *Topographic surveys* map the relief of a local land surface (and the objects thereon). Such measurements may be made either with traditional surveying instruments or with the use of aerial photographs (photogrammetry).

- *Route surveys* involve the development and control of public or private improvement projects, such as highways, pipelines, and so on.

- *Construction staking* involves the field controls used during construction of improvement projects.

B. Survey Party

In general, the following personnel and equipment comprise a *survey party*. However, many organizations use different combinations of personnel and equipment.

Personnel

- The *party chief* is responsible for all aspects of the field work including personnel assignments and training, equipment, and scheduling. Often, the party chief will perform in another capacity as well, such as operating the instrument as the instrumentperson, or as the note keeper.

- The *instrumentperson* is responsible for setting up and operating the field equipment. Often, the instrumentperson will function as note keeper as well.

- The *rodperson* holds the surveying rod plumb at locations indicated by the instrumentperson.

- The *chainpersons* (who always operate in pairs) stretch a steel measuring tape between them to determine lineal distance.

- The *stake setter*, also nicknamed the *ginney hopper*, drives wooden stakes into the ground at locations indicated by either the instrumentperson or the rodperson.

Equipment

- A *transit* measures horizontal and vertical angles using a calibrated mechanical plane and a telescope.

- A *level* measures differences in elevation between points on the ground. More accurately, it measures the difference in elevation between the instrument itself and a point on the ground. A graduated rod is used in conjunction with the level.

- *Electronic distance measurement* (EDM) devices consist of a laser emitter mounted on the instrument and a prism mounted on a rod. The emitter sends a beam of light to the prism, which reflects the beam back to the emitter. The change in wavelength between the emitted and reflected beam is factored to yield a distance measurement.

- A *total station* is an EDM that measures both horizontal and vertical angles and distances.

- *Stadia* is the term used to describe a particular method of determining distances indirectly. A telescope with unique crosshairs is used with a specially graduated rod.

2. Engineering Surveying in California Law

A. California Laws

Under California state law, licensed land surveyors are authorized to practice all types of surveying. Licensed civil engineers, however, can lawfully practice only some of these same activities.

As a general concept, civil engineers cannot practice *boundary (property line) surveying*, but are authorized to practice the other types of land surveying provided that the survey work is related to a public or private improvement project. That is, civil engineers can take measurements relating to the position of such facilities as roadways, buildings, pipelines, and so on, but cannot establish property lines. The following sections outline the Professional Land Surveyors Act, the Professional Engineers Act, and the Subdivision Map Act.

B. Professional Land Surveyors Act

Land surveying in California is regulated by the Business and Professions Code, Sections 8700 to 8806, known as the *Professional Land Surveyors Act*. Section 8725 of this act defines the requirement for a license.

Any person practicing or offering to practice land surveying in this state shall submit evidence that he or she is qualified to practice and shall be licensed under this chapter.

It is unlawful for any person to practice, offer to practice, or represent himself or herself as a land surveyor in this state, or to set, reset, replace, or remove any survey monument on land in which he or she has no legal interest, unless he or she has been licensed or *specifically exempted* from licensing under this chapter.

Section 8731 exempts civil engineers registered prior to January 1, 1982, from license requirement.

A registered civil engineer is exempt from licensing under this chapter and may engage in the practice of land surveying with the same rights and privileges and the same duties and responsibilities of a licensed land surveyor, provided that for civil engineers who become registered after January 1, 1982, they shall pass the second division examination provided for in Section 8741, and obtain a land surveyor's license, before practicing land surveying as defined in this chapter.

Section 8726 defines a land surveyor as

... a person, including any person employed by the state or by a city, county, or city and county within the state, who practices land surveying within the meaning of this chapter who, either in a public or private capacity, does or offers to do any one or more of the following:

(a) Locates, relocates, establishes, reestablishes, or retraces the alignment or elevation for any of the fixed works embraced within the practice of civil engineering, as described in Section 6731.

(b) Determines the configuration or contour of the earth's surface, or the position of fixed objects thereon or related thereto, by means of measuring lines and angles, and applying the principles of mathematics or photogrammetry.

(c) Locates, relocates, establishes, reestablishes, or retraces any property line or boundary of any parcel of land, right-of-way, easement, or alignment of those lines or boundaries.

(d) Makes any survey for the subdivision or resubdivision of any tract of land. For the purposes of this subdivision, the term "subdivision" or "resubdivision" shall be defined to include, but not limited to, the definition of the Subdivision Map Act.

(e) By the use of the principles of land surveying determines the position for any monument or reference point which marks a property line, boundary, or corner, or sets, resets, or replaces any such monument or reference point.

(f) Geodetic or cadastral surveying.

(g) Determines the information shown or to be shown on any map or document prepared or furnished in connection with any one or more of the functions described in subdivisions (a), (b), (c), (d), (e), and (f).

(h) Indicates, in any capacity or in any manner, by the use of the title land surveyor or by any other title or by any other representation that he or she practices or offers to practice land surveying in any of its branches.

(i) Procures or offers to procure land surveying work for himself, herself, or others.

(j) Manages, or conducts as manager, proprietor, or agent, any place of business from which land surveying work is solicited, performed, or practiced.

(k) Coordinates the work of professional, technical, or special consultants in connection with the activities authorized by this chapter.

(l) Determines the information shown or to be shown within the description of any deed, trust deed, or other title document prepared for the purpose of describing the limit of real property in connection with any one or more of the functions described in subdivisions (a) to (f), inclusive.

(m) Creates, prepares, or modifies electronic or computerized data in the performance of the activities described in subdivisions (a), (b), (c), (d), (e), (f), (k), and (l).

Any department or agency of the state or any city, county, or city and county which has an unregistered person in responsible charge of land surveying work on January 1, 1986, shall be exempt from the requirement that the person be licensed as a land surveyor until such time as the person currently in responsible charge is replaced.

The review, approval, or examination by a governmental entity of documents prepared or performed pursuant to this section shall be done by, or under the direct supervision of, a person authorized to practice land surveying.

C. Professional Engineers Act

Civil engineers registered after January 1, 1982, are authorized to perform those activities shown in Section 8726 (a), (b), and (m). This authorization is granted by virtue of Section 6731 of the Business and Professions Code. (Sections 6700 through 6799 are known as the Professional Engineers Act.)

Civil engineering includes the practice or offer to practice, either in a public or private capacity, the following:

(a) Locates, relocates, establishes, reestablishes, or retraces the alignment or elevation for any of the fixed works embraced within the practice of civil engineering.

(b) Determines the configuration or contour of the earth's surface or the position of fixed objects thereon or related thereto, by means of measuring lines and angles, and applying the principles of trigonometry and photogrammetry.

(c) Creates, prepares, or modifies electronic or computerized data in the performance of the activities described in subdivisions (a) and (b).

Civil engineering embraces the following studies or activities in connection with fixed works for irrigation, drainage, waterpower, water supply, flood control, inland waterways, harbors, municipal improvements, railroads, highways, tunnels, airports and airways, purification of water, sewerage, refuse disposal, foundations grading, framed and homogeneous structures, buildings or bridges:

(a) The economics of, the use and design of, materials of constructions and the determination of their physical qualities.

(b) The supervision of the construction of engineering structures.

(c) The investigation of the laws, phenomena and forces of nature.

(d) Appraisals or valuations.

(e) The preparation or submission of designs, plans and specifications and engineering reports.

(f) Coordination of the work of professional, technical, or special consultants.

(g) Creation, preparation, or modification of electronic or computerized data in the performance of the activities described in subdivisions (a) through (f).

Civil engineering also includes city and regional planning insofar as any of the above features are concerned therein.

Although not statutorily defined, the term *engineering surveying* has been accepted to mean topographic, construction, and alignment surveying.

D. Subdivision Map Act

The Subdivision Map Act is the California state law governing the procedures by which *boundary surveys* are officially recorded by the local municipality in which the property is located. Examinees in the land surveyor licensing examination are tested extensively on the Map Act. However, since boundary surveying is beyond the authority of registered civil engineers in California, examinees in the special civil surveying examination are not tested on this law. A basic knowledge of the subject is valuable, though, when answering questions of authority to practice.

Essentially, the Map Act requires that boundary surveys fall into one of two map types.

■ A *parcel map* is used where a large parcel is subdivided into four or fewer smaller ones.

■ A *subdivision map* (or *final map*) is used where a large parcel is subdivided into five or more smaller ones.

Each of the map types show only property lines; no indications of roadways, buried utility lines, and so on, are to be included. In addition, each of the maps will ultimately (after a lengthy approval process) be officially recorded by the county recorder.

A developer is required to submit to the local municipality an application for one of the three types of maps for each project. Such applications must be accepted as proper by the municipality before the municipality considers the proposed boundary changes. For example, if a developer files a parcel map application showing a large parcel being subdivided into seven smaller ones, the municipality will return the application with an indication that the project should be resubmitted as a subdivision map application. Once the developer has resubmitted the project, the larger issues concerning zoning and required public improvements will be addressed.

As part of any subdivision map application, and as part of some parcel map applications, the municipality can require a *tentative map*. Such maps are preliminary versions of either the subdivision or parcel map. As such, no license of any kind is required for individuals preparing tentative maps.

The Map Act mainly addresses the level of authority granted to the municipality to require, as a condition of approval, certain public improvements from developers. Such improvements include roadways, storm and sanitary sewage facilities, buildings, water systems, waste treatment facilities, and so on. The Map Act requires that all conditions of approval as negotiated with the municipality be fully satisfied before the map is officially recorded.

If a subdivision lies within a city, approval of the final or parcel map may be the responsibility of the city engineer. If that person is a post-1982 registered civil engineer, an additional signature is required to certify that the map is technically correct. This certification must be made by a person authorized to practice land surveying. Section 66442 of the Map Act contains the requirements for a city engineer's statement on a final map.

(a) If a subdivision for which a final map is required lies within an unincorporated area, a certificate or statement by the county surveyor is required. If a subdivision lies within a city, a certificate or statement by the city engineer or city surveyor is required. The appropriate official shall sign, date, and, below or immediately adjacent to the signature, indicate his or her registration or license number with expiration date and state that:

(1) He or she has examined the map.

(2) The subdivision as shown is substantially the same as it appeared on the tentative map, and any approved alterations thereof.

(3) All provisions of this chapter and of any local ordinances applicable at the time of approval of the tentative map have been complied with.

(4) He or she is satisfied that the map is technically correct.

(b) City or county engineers registered as civil engineers after January 1, 1982, shall only be qualified to certify the statements of paragraphs (1), (2), and (3) of subdivision (a). The statement specified in paragraph (4) shall only be certified by a person authorized to practice land surveying pursuant to the Professional Land Surveyors' Act or a person registered as a civil engineer prior to January 1, 1982. The county surveyor, the city surveyor, or the city engineer, as the case may be, or other public official or employee qualified and authorized to perform the functions of one of those officials, shall complete and file with his or her legislative body his or her certificate or statement, as required by this section, within twenty days from the time the final map is submitted to him or her by the subdivider for approval.

Section 66450, pertaining to Parcel Maps, contains the same requirements.

Practice Problems

1. Which of the following surveys documents property lines?

    (A)   route survey

    (B)   topographic survey

    (C)   geodetic survey

    (D)   boundary survey

2. Which of the following may post-1982 registered civil engineers do?

    (A)   perform boundary surveys

    (B)   prepare a parcel map

    (C)   lay out a shopping center

    (D)   prepare land descriptions for new easement deeds

3. Prior to performing a topographic survey on a large parcel of land, the surveyors observe two fence lines along one side of the property, very near and almost parallel with each other. Prior to the start of the project, which of the following should they do?

    (A)   ask the owner which fence is his or hers

    (B)   establish the property line and locate the fences relative to the line

    (C)   secure the services of a person authorized to practice land surveying

    (D)   do the topo survey from an assumed horizontal datum—the fences are not the surveyors' problem

4. Which of the following is true for post-1982 registered civil engineers?

    (A)   They may not become city engineers.

    (B)   They may become city engineers and have the authority to state that a subdivision map complies with an approved tentative map.

    (C)   They may become city engineers and approve subdivision maps.

    (D)   They may become city engineers and state that a subdivision map is technically correct.

Solutions

1.   Answer (D)

2.   Answer (C)

3.   Answer (C)

4.   Answer (B)

Horizontal Distance Measurement 2

1. Introduction

Horizontal distances can be measured by chaining (or taping), stadia, and electronic distance measuring instruments (EDM). All of the methods, except stadia, are *direct measurement* techniques. Stadia measures parameters related to distance, which are later converted to distances.

2. Horizontal Distances

Since most land is not flat, surveyors generally measure along the sloped ground and later use trigonometry to reduce the measurements into horizontal and vertical components (Fig. 2.1(a)). If the surveyor wants to measure the difference in horizontal and vertical quantities between two points, the same method can be used (Fig. 2.1(b)).

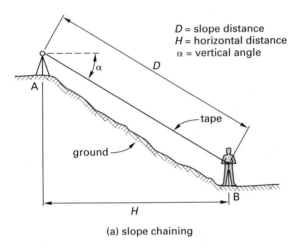

D = slope distance
H = horizontal distance
α = vertical angle

(a) slope chaining

D = slope distance
H = horizontal distance
V = difference in elevation

(b) chaining out of level

***Figure 2.1** Measuring a Slope Distance*

In Fig. 2.1, the horizontal distance, H, is computed using Eq. 2.1.

$$H = D \cos \alpha \qquad 2.1$$

3. Taping or Chaining

A. Introduction

Taping or *chaining* involves two people, each positioned at points of interest, measuring the lineal distance between them by using a graduated steel tape. The predecessor of the modern steel tape was a link chain calibrated at 66 ft long, called a *Gunter's chain.*

The two chainpersons are designated as either *forward* or *rear* when measuring distances along an *alignment line.* The person up station from the other is the forward chainperson, while the individual positioned down station is the rear chainperson. Such measurements are said to be *on line.*

The two chainpersons should strive to hold their respective tape ends level. By using a device known as the *hand level*, the chainpersons can determine the difference in elevation between them and hold the tape accordingly. If the lower chainperson must hold the chain higher than shoulder height, a *plumb-bob* is used to position the tape over the point.

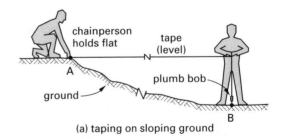

(a) taping on sloping ground

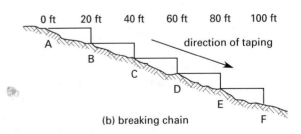

(b) breaking chain

***Figure 2.2** Chaining*

In Fig. 2.2(a), the lower chainperson determines a point where his or her shoulders are level with the high point and holds the tape at that level. The tape ends must be level: A difference of 2.00 ft in 100.00 ft will cause an error of 0.02 ft.

Figure 2.2(b) represents a method of chaining down a long slope in increments. The example shows 20.00 ft increments, but this can be changed based on the steepness of the slope. To start, the upper chainperson holds 0.00 at point A and the lower chainperson sets at point B on line at 20.00 ft. Then the chainpersons move downhill, the upper chainperson holding the 20.00 ft mark at B, and the lower chainperson setting at point C on line at 40.00 ft. The process continues until the 100.00 ft point at F is set (unless a tape longer than 100.00 ft is used). As in Fig. 2.2(a), the lower chainperson should set to hold the tape level and at shoulder height.

B. Units of Measurement—Chaining

The earlier surveys in the United States were made with a Gunter's chain (named after its inventor, Edmund Gunter). This four-pole chain is 66.00 ft long and is made of 100 links, each 0.66 ft long (7.92 in). The chain is still used by the Bureau of Land Management (BLM) as the official unit of length when making public land surveys.

The unit of measurement used today is the U.S. survey foot.

Table 2.1 *Units of Measurement*

1 pole = 16.5 feet

1 chain = 4 poles = 66.00 feet = 100 links

1 link = 0.66 feet

1 mile = 80 chains = 320 rods = 5280 feet

1 acre = 10 square chains

Table 2.2 *Units of Linear Measurement*

unit	inches	feet	yards	meters
1 in	1	0.08333	0.02778	0.0254
1 ft	12	1	0.333	0.3048
1 yd	36	3	1	0.9144
1 m	39.37	3.2808333	1.094	1

The Gunter's chain was replaced by the steel tape in the late 1890s. Steel tapes vary in width from $1/4$ in to $5/16$ in and in thickness from 0.008 in to 0.025 in. The 100.00 ft tape is the most common tape used, although the 200.00 ft and 300.00 ft tapes have their applications.

The three most common types of tape are the add tape, the cut tape, and the fully graduated tape.

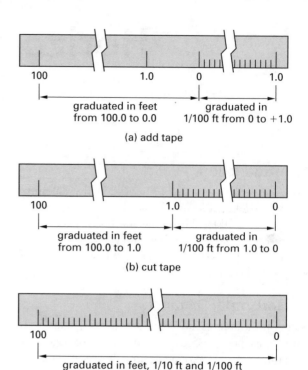

Figure 2.3 *Steel Tapes*

In Fig. 2.3(a), the rear chainperson reads an even foot at his or her position, and the forward chainperson reads the *add dimension* at hers or his. The distance measured is the even foot plus the add dimension.

In Fig. 2.3(b), the rear chainperson reads an even foot, the forward chainperson reads the *cut dimension*, and the note keeper subtracts the cut dimension from the even foot call. This result is noted as the measured distance.

To use the fully graduated tape (Fig. 2.3(c)), the rear chainperson holds zero and the forward chainperson reads the measured distance directly from the tape.

C. Temperature Correction

Since tapes are made of steel, they are affected by changes in air temperature—that is, they expand when it is hot and contract when it is cold. Tapes are standardized at 68°F. A change of 15°F will make a change of 0.01 ft in a 100.00 ft tape. The exact coefficient of expansion for steel is 0.00000645 per degree Fahrenheit.

The formula for the amount of correction to be applied to a measured length using a 100.00 ft steel tape is

$$c = \left(0.00000645 \; \frac{1}{°F}\right)(T - T_o)(100 \text{ ft}) \qquad 2.2$$

c is the correction in feet, T is the temperature of the tape, and T_o is the standardized temperature (assume 68°F unless stated otherwise).

Example 2.1

A line is measured as 795.83 ft at 29°F. Using a standard tape, what is the true length of the line?

Solution

$$c = \left(0.00000645\ \frac{1}{°F}\right)(29°F - 68°F)(795.83\ \text{ft})$$

$$= -0.20\ \text{ft}$$

$$L = 795.83\ \text{ft} - 0.20\ \text{ft} = 795.63\ \text{ft}$$

Example 2.2

A line is measured as 800.96 ft at 98°F. Using a standard tape, what is the true length of the line?

Solution

$$c = \left(0.00000645\ \frac{1}{°F}\right)(98°F - 68°F)(800.96\ \text{ft})$$

$$= +0.15\ \text{ft}$$

$$L = 800.96\ \text{ft} + 0.15\ \text{ft} = 801.11\ \text{ft}$$

Note that when a tape is cold, the correction to the measured distance is subtracted, and when the tape is hot, the correction is added.

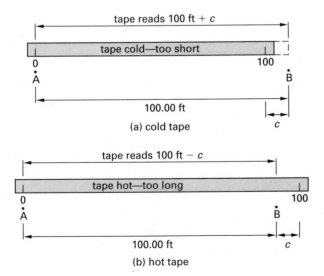

Figure 2.4 *Effects of Measuring Between Points A and B with a Cold Tape and a Hot Tape*

D. Proper Tension

The standard tension applied to a 100.00 ft steel tape that is fully supported (lying flat on the ground) is usually 10 lbf. The standard tension applied to the same tape supported at the ends only is 30 lbf. A decrease or increase of 15 lbf will cause the length to vary less than 0.01 ft. As tapes wear out and become smaller and lighter, the required tension changes accordingly. Tapes must be calibrated at regular intervals. Many agencies have an area set aside for this purpose, where two points are established at a known distance, usually 100.00 ft apart. The temperature is measured and a correction is applied to the tape to determine what distance should be read. Then the tension is adjusted to obtain the correct reading and the tension used is noted. Usually, a brass tag is attached to the tape and is marked with the tension and temperature that will yield 100.00 ft.

4. Stadia Measurements

Horizontal and vertical distances can be measured indirectly by reading the intercept distance measured using the off-center crosshairs on the lens of a transit together with a rod. There are four crosshairs on the telescope lens in Fig. 2.5.

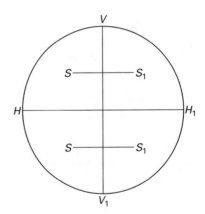

Figure 2.5 *Stadia Lens Crosshairs*

V-V_1 is the vertical crosshair, H-H_1 is the horizontal crosshair, and S-S_1 are the *stadia hairs*. The stadia hairs are graduated such that when the scope is horizontal and the rod is vertical, the difference between the upper and lower rod reading(s) can be factored to yield the horizontal distance from the instrument to the rod. The factors used are equipment dependent, and are shown in Eq. 2.3.

$$\text{distance} = KS + C \qquad 2.3$$

Typical values of K are 100, 200, and 333. Values of C are 0 ft for internal-focusing equipment and 1 ft for external-focusing. Values of both K and C must be given.

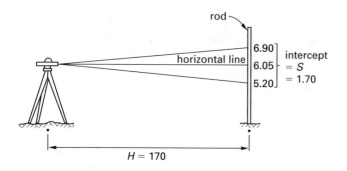

Figure 2.6 *Horizontal Distance by Stadia*

Alternatively, slope measurements using stadia methods must be reduced to horizontal distances. When an *inclined stadia measurement* is made, the rod is either held perpendicular to the ground or held plumb, thus yielding a greater intercept than if it were held perpendicular to the line of sight. Two corrections must be made to compute the horizontal distance, as illustrated in Fig. 2.7.

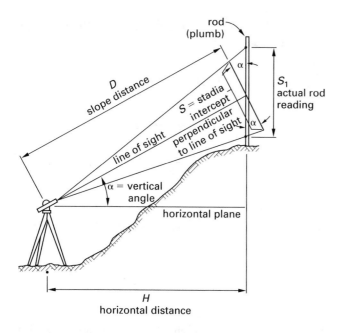

Figure 2.7 *Inclined Stadia Measurement*

The formulae for inclined stadia reduction are

$$S = S_1 \cos \alpha \qquad 2.4$$

$$D = 100S = 100S_1 \cos \alpha \qquad 2.5$$

$$H = D \cos \alpha = 100S_1 \cos^2 \alpha \qquad 2.6$$

As an aid to reducing inclined stadia measurements, tables for the value $100 \cos^2 \alpha$ have been developed and are included in App. A.

Example 2.3

A stadia intercept of 4.35 is read on the rod, and the vertical angle is $5°10'$. Use Eq. 2.6 to find the horizontal distance, and then check your answers using the tables in App. A.

Solution

$$\cos \alpha = 0.995937$$
$$\cos^2 \alpha = 0.991890$$
$$H = 100S_1 \cos^2 \alpha$$
$$= (100)(4.35 \text{ ft})(0.991890)$$
$$= 431.47 \text{ ft}$$

From the tables in App. A, the horizontal distance per 100.00 ft for $5°10'$ is 99.19 ft, and $H = (4.35 \text{ ft})(99.14) = 431.48$ ft.

5. Electronic Distance Measuring Instruments

A. Federal Principles

Electronic distance measuring instruments, EDMs, using lightwaves, were invented in the 1940s. The microwave type were developed about 10 years later. The large, heavy microwave instruments, although more expensive than the lightwave devices, measured very long distances. However, the need for two operators made the microwave system less desirable, giving way to the smaller EDM using a reflector system.

The basic principle behind the EDM is that it measures wavelengths of a lightwave as it travels to a reflector and is sent back to the emitting unit.

Signals need to be corrected for atmospheric pressure and temperature. The errors caused by varying atmospheric pressure and temperature are very small and are measured in parts per million. One part per million in a 2000 m (6500 ft) distance would be only 2 mm.

The correct constant must be used for the type of reflector used with an EDM. The lightwave travels a fixed distance throughout the prism. This distance must be subtracted from the measured distance by setting a prism offset constant in the EDM that automatically subtracts this amount so that the displayed distance is the corrected one.

Most prisms today have a 30 mm or 40 mm offset. Using the wrong constant could result in an error of 0.03 ft in every distance measured.

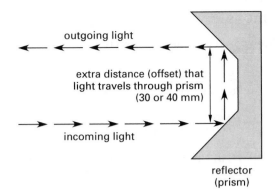

Figure 2.8 *Prism Offset*

B. Total Stations

A *total station* is an EDM that measures both distances and angles. Most of these units display the following:

- slope distance

- horizontal distance

- vertical distance (difference in elevation between the HI of the prism and the instrument)

- vertical angle

- horizontal angle

Some total stations have an onboard computer that will give x-, y-, and z-coordinates and compute direction and distance to predetermined stations. Most total stations have data collector capability. When a data collection device is attached to the EDM, horizontal angles and distances, vertical angles and distances, and point descriptions are automatically captured, stored, and dumped into the office computer system. A plot or map can be made automatically without any note taking, note reduction, or manual plotting.

Practice Problems

1. A slope distance of 87.52 ft is measured between points A and B. The zenith angle is measured as 72°25′40″. The tape is standardized at 68°F, and its temperature is 92°F. What is the horizontal distance from A to B?

- (A) 83.43 ft
- (B) 83.44 ft
- (C) 83.45 ft
- (D) 83.46 ft

2. A slope distance of 93.86 ft is measured directly between two points, C and D. The elevation of C is 125.65 ft, and the elevation of D is 92.40 ft. What is the horizontal distance CD? There is no temperature correction.

- (A) 87.77 ft
- (B) 87.80 ft
- (C) 87.82 ft
- (D) 87.97 ft

3. A horizontal distance is measured as 897.45 ft with a tape standardized at 60°F. If its temperature is 97°F, what is the corrected distance?

- (A) 897.28 ft
- (B) 897.34 ft
- (C) 897.62 ft
- (D) 897.66 ft

4. A slope distance of 1110.60 ft is measured between two monuments along a sloping street. The difference in elevation between the two monuments is 37.60 ft, and the temperature is 33°F. If the tape is standardized at 68°F, what is the corrected horizontal distance between the two monuments?

- (A) 1109.61 ft
- (B) 1109.71 ft
- (C) 1109.96 ft
- (D) 1110.21 ft

5. How long is a Gunter's chain?
- (A) 100 links
- (B) 66.00 ft
- (C) 4 poles
- (D) all of the above

6. A distance of 372.87 chains equals which of the following?
- (A) 7502 m
- (B) 24,609.42 ft
- (C) 8200.14 yd
- (D) all of the above

7. A parcel of land measures 17.3 chains by 9.7 chains. How many acres does it contain?
- (A) 16.781
- (B) 17.681
- (C) 168.71
- (D) 176.81

8. 1 m is equivalent to which of the following?

   (A) 39.37 in

   (B) 3.2808333 ft

   (C) 1.094 yd

   (D) all of the above

9. Which of the following is true on an add tape?

   (A) Every foot, tenth, and hundredth is graduated.

   (B) There are graduations every hundredth of a foot from the 1.0 ft mark to the 0.0 ft mark.

   (C) There are graduations from the 0.0 ft mark in hundreds to 1.0 ft beyond the 0.0 ft mark.

   (D) None of the above is true.

10. A tape standardized at 68°F is used to set two points, 1675.25 ft apart. The temperature is 40°F. What distance should the chairpersons measure to set the correct distance?

   (A) 1674.95 ft

   (B) 1675.17 ft

   (C) 1675.33 ft

   (D) 1675.55 ft

11. An instrument is set over point A at an HI of 5.1 ft. An inclined stadia measurement to point B is made. The stadia hair intercepts are upper hair—7.25 ft, middle hair—5.10 ft, and lower hair—2.95 ft. The vertical angle to the rod is measured at 22°39′. What is the horizontal distance AB?

   (A) 366.23 ft

   (B) 367.18 ft

   (C) 369.42 ft

   (D) 400.00 ft

12. A correction for atmospheric conditions and temperature made to the distance read on an EDM is 3 ppm. In 5000 ft, this equates to which of the following?

   (A) 0.005 yd

   (B) 4.6 mm

   (C) 0.015 ft

   (D) all of the above

13. For which of the following are EDMs best suited?

   (A) for construction surveys

   (B) for topographic surveys

   (C) for control surveys

   (D) for all of the above

Solutions

1. Apply a temperature correction to the slope distance.

$$c = \left(0.00000645 \; \frac{1}{°F}\right)(T - T_o)(87.52 \text{ ft})$$

$$= \left(0.00000645 \; \frac{1}{°F}\right)(92°F - 68°F)(87.52 \text{ ft})$$

$$= +0.014 \text{ ft} = +0.01 \text{ ft} \quad [\text{tape hot, too long}]$$

$$D = 87.52 \text{ ft} + 0.01 \text{ ft} = 87.53 \text{ ft}$$

Using the corrected slope distance, compute the horizontal distance.

$$H = D \sin Z$$

$$= (87.53 \text{ ft})(\sin 72°25'40'')$$

$$= 83.446 \text{ ft} \quad (83.45 \text{ ft})$$

Answer (C)

2. In the given problem,

$$D = 93.86 \text{ ft}$$

$$V = 125.65 \text{ ft} - 92.40 \text{ ft} = 33.25 \text{ ft}$$

Using Eq. 2.2, solve for H.

$$H = \sqrt{D^2 - V^2}$$

$$= \sqrt{(93.86 \text{ ft})^2 - (33.25 \text{ ft})^2}$$

$$= 87.77 \text{ ft}$$

Answer (A)

3. Using Eq. 2.3, compute the temperature correction.

$$c = \left(0.00000645 \; \frac{1}{°F}\right)(T - T_o)(897.45 \text{ ft})$$

$$= \left(0.00000645 \; \frac{1}{°F}\right)(97°F - 60°F)(897.45 \text{ ft})$$

$$= +0.21 \text{ ft} \quad [\text{tape hot, too long}]$$

$$D = 897.45 \text{ ft} + 0.21 \text{ ft} = 897.66 \text{ ft}$$

Answer (D)

4. Apply the temperature correction to the slope distance.

$$c = \left(0.00000645 \; \frac{1}{°F}\right)(T - T_o)(1110.60 \text{ ft})$$

$$= \left(0.00000645 \; \frac{1}{°F}\right)(33°F - 68°F)(1110.60 \text{ ft})$$

$$= -0.25 \text{ ft} \quad [\text{tape cold, too short}]$$

$$D = 1110.60 \text{ ft} - 0.25 \text{ ft} = 1110.35 \text{ ft}$$

Using the corrected slope distance, compute the horizontal distance.

$$D = 1110.35 \text{ ft}$$
$$V = 37.60 \text{ ft}$$
$$H = \sqrt{D^2 - V^2}$$
$$= \sqrt{(1110.35 \text{ ft})^2 - (37.60 \text{ ft})^2}$$
$$= 1109.71 \text{ ft}$$

Answer (B)

5. Answer (D)

6. $(372.87 \text{ chains})\left(66 \dfrac{\text{ft}}{\text{chain}}\right) = 24{,}609.42 \text{ ft}$

Answer (B)

7. $\dfrac{(17.3 \text{ chains})(9.7 \text{ chains})}{\dfrac{10 \ (\text{chains})^2}{\text{ac}}} = \dfrac{167.81 \ (\text{chains})^2}{\dfrac{10 \ (\text{chains})^2}{\text{ac}}}$

$$= 16.718 \text{ ac}$$

Answer (A)

8. Answer (D)

9. Answer (C)

10. Compute the correction using Eq. 2.4.

$$c = \left(0.00000645 \ \frac{1}{{}^\circ\text{F}}\right)(T - T_o)(1675.25 \text{ ft})$$
$$= \left(0.00000645 \ \frac{1}{{}^\circ\text{F}}\right)(40{}^\circ\text{F} - 68{}^\circ\text{F})(1675.25 \text{ ft})$$
$$= -0.30 \text{ ft}$$

The tape is cold and too long. To set a distance, reverse the sign of the correction and apply it to the desired distance. Therefore, the chainpersons should measure a distance of

$$D = 1675.25 \text{ ft} + 0.30 \text{ ft} = 1675.55 \text{ ft}$$

Answer (D)

11. Subtract the lower reading from the upper reading to compute S_1.

$$S_1 = 7.25 \text{ ft} - 2.95 \text{ ft} = 4.30 \text{ ft}$$

Using Eq. 2.6,

$$H = (100)(S_1)\cos^2 \alpha$$
$$= (100)(4.30 \text{ ft})(\cos^2 22{}^\circ 39')$$
$$= 366.23 \text{ ft}$$

Answer (A)

Alternate solution: Using the stadia table in App. A, find the reduction factor for $22{}^\circ 39'$ by interpolation.

$22{}^\circ 38' = 85.19$
$22{}^\circ 40' = 85.15$
difference $= 0.04$
divide by $2 = 0.02$
add to $22{}^\circ 40' = 85.15 + 0.02 = 85.17$
$H = (85.17)(4.30 \text{ ft}) = 366.23 \text{ ft}$

Answer (A)

12. $\quad\quad \dfrac{3}{1 \times 10^6} = \dfrac{x \text{ ft}}{5000 \text{ ft}}$

$$x = 0.015 \text{ ft}$$

$$(0.015 \text{ ft})\left(0.3048 \ \frac{\text{m}}{\text{ft}}\right) = 0.0046 \text{ m} = 4.6 \text{ mm}$$

$$\dfrac{0.015 \text{ ft}}{3 \ \dfrac{\text{ft}}{\text{yd}}} = 0.005 \text{ yd}$$

Answer (D)

13. Answer (D)

Vertical Distance Measurement 3

Contributed by John D. Pavlik, PLS

1. Introduction

The *elevation* of a point is the vertical distance above or below a given reference or datum. The most commonly used datum in the United Status is Mean Sea Level–1929 Adjustment (MSL29). *Leveling* is a method by which the elevation of points on the ground, or on or within structures, are measured either directly or indirectly.

2. Direct Leveling

A. Leveling Circuit

A *leveling circuit* (or *bench circuit*) uses a surveyor's level, which is a telescope (with crosshairs on the lens) mounted on a planar surface. The planar surface is adjusted during setup such that it is horizontal (parallel to a horizontal datum) using a spirit (or bubble) level mounted on the instrument. The telescope movement is thus restricted to rotation about the horizontal plane.

After setting up the level, the next step is to determine the elevation of the instrument (telescope) itself. By placing a graduated rod onto a point of known elevation such as a benchmark, the *rod reading* (RR) at this point is added to the known elevation to yield the *height of instrument* (HI). Then, to determine the elevation of any point of interest, the rod is placed at a point (such as a future building corner or centerline of a highway) and the rod reading there is subtracted from the HI. When taking rod readings of, say, two points along an alignment line, a rod reading taken on the point down station of the instrument is often termed the *backsight* (BS), while a rod reading taken on the point up station is the *foresight* (FS).

The term *benchmark* refers to a monument of known elevation that is usually informally filed with local agencies. A benchmark monument is often called out as a point on traverses (which involves only horizontal quantities), but strictly speaking a benchmark is a point of vertical control only.

Temporary benchmarks (TBMs) are those points of newly determined elevation marked by either a stake or a chiseled corner of a concrete curb or other convenient location. The elevation and location of TBMs is not filed with the local agencies; they are intended for use by only the individual placing the TBM.

A *turning point* (TP) is a point of newly determined elevation used to determine the HI at an instrument's new setup location. Turning points are used when the initial setup location is too far away from some of the points of interest in the bench circuit. While the instrumentperson is able to focus on the rod at points in the circuit close to the setup point, the instrument must be relocated to a second setup point to effectively take rod readings on those points farther away. To determine the new HI at the second setup point, one of the closer newly determined points is used as the point of known elevation, much as a benchmark is used.

Example 3.1

What is the elevation of point Y?

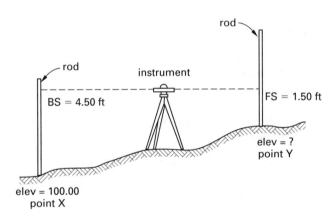

Solution

$$
\begin{array}{r}
\text{elev at X} = 100.00 \text{ ft} \\
\text{BS} = +4.50 \text{ ft} \\
\hline
\text{HI} = 104.50 \text{ ft} \\
\text{FS} = -1.50 \text{ ft} \\
\hline
\text{elev at Y} = 103.00 \text{ ft}
\end{array}
$$

This setup is repeated many times with the rodperson and instrumentperson moving forward to establish elevations using turning points, temporary benchmarks, and benchmarks.

The instrument should always be set up so the distance to the backsight is equal to the distance to the foresight. This eliminates the correction for curvature and refraction.

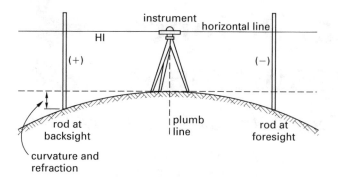

Figure 3.1 Equal Foresight and Backsight

The most common procedural errors in leveling are the following:

- The instrument is out of adjustment.

- The rod is not standard length.

- The rod is not held plumb.

- There are mistakes in rod readings.

B. Precise Leveling

Precise leveling is spirit leveling of a high order of accuracy, usually extended over large areas, to furnish accurate vertical control (new benchmarks) as a basis of control for lower-order work. More precise equipment is used to accomplish the measurement of vertical distances directly.

Precise levels have more sensitive level tubes, improved optics, and greater magnification power, and they are usually equipped with a micrometer. Precise leveling rods have an invar steel tape kept under constant tension on the face. Invar has a very low coefficient of expansion, and the rods are calibrated to an exact length.

3. Indirect Leveling

Trigonometric leveling indirectly determines the elevation of a point by measuring vertical angles and horizontal or slope distances.

Example 3.2

Determine the difference in elevation between X and Y given the vertical angle α, the slope distance D, a rod reading (RR), a height of instrument (HI), and an elevation of X = 303.00 ft.

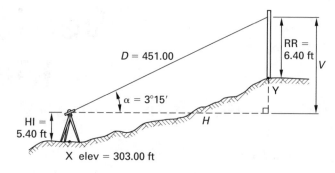

Solution

$$V = D \sin \alpha$$
$$= (415.00 \text{ ft})(\sin 3°15')$$
$$= 25.57 \text{ ft}$$

$$\text{elev at instrument} + \text{HI} \pm V - \text{RR} = \text{elev at rod}$$
$$303.00 \text{ ft} + 5.40 \text{ ft} + 25.57 \text{ ft} - 6.40 \text{ ft} = 327.57 \text{ ft}$$

The slope distance was measured using *stadia*, a method of tachometry by which horizontal distances are determined indirectly using subtended rod intervals.

4. Leveling Rods

Leveling rods are made of wood, metal, or fiberglass and are graduated in feet or meters. The lower end of the rod is usually the point of zero measurement and is shod with metal to protect it. There are two general classes of rods: *self-reading* and *self-reducing* rods.

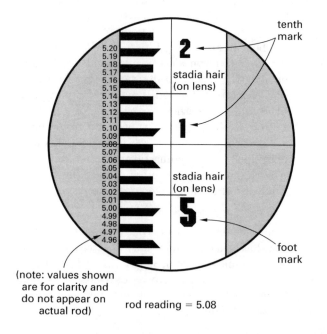

Figure 3.2 Philadelphia Rod

The most widely used self-reading rod is the *Philadelphia rod*. It is made in two sliding sections that can be extended to 13 ft. It is held vertically (plumb) on a point, and the crosshair of the level intercepts the rod at the vertical distance above the point.

The *Lenker rod* is a self-reducing rod that has a continuously graduated face that can be moved such that the elevation of the point can be set at the intercept of the crosshair. No notes need to be taken. The values on the rod increase from top to bottom because the higher the elevation of the point, the higher the reading on the rod.

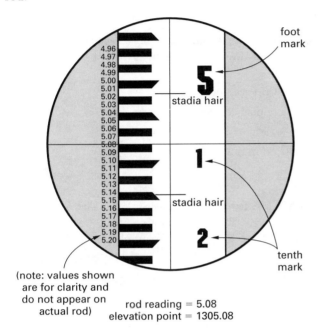

(note: values shown are for clarity and do not appear on actual rod)

rod reading = 5.08
elevation point = 1305.08

Figure 3.3 *Lenker Rod*

5. Profile and Cross-Section Leveling

Profile leveling involves obtaining elevations along a survey line or centerline at given increments and grade breaks. The procedure usually consists of setting up the level, establishing an HI, and taking a series of sideshots (SS) or intermediate foresights before turning ahead. Some sample notes are shown in Table 3.1.

Cross-sectioning involves obtaining elevations at the grade breaks along a typical cross section that is at right angles to the centerline.

6. Reciprocal Leveling

When establishing new elevations using differential leveling, it is advisable to balance backsights and foresights. When this cannot be done, reciprocal leveling is used. See Fig. 3.4.

Setups 1 and 2 should be performed with minimal time between them. The difference in elevation is computed by using the mean (average) of the backsights and the mean of the foresights. The short distances from the setup point to the turning point should be nearly equal. A good distance to use is between 15 ft and 20 ft. Table 3.2 is an example of how a *turn* (same function as a turning point) using reciprocal leveling is accomplished.

Table 3.1 *Cross-Section Notes*

centerline profile of Rio Grande Avenue		1/31/94				C. Landmann–Inst.
				68°F clear		P. Cuomo–Rod
sta.	BS	HI	SS	FS	elev	
BM 1	3.25	215.35			212.10	3 in alum. disk stamped "County Surveyor" at SE corner of Rio Grande and First
10+00			4.00		211.35	
10+50			4.25		211.10	
11+00			4.50		210.85	
11+50			4.75		210.60	
12+00			5.00		210.35	
TP 1	3.50	213.75		5.10	210.25	
12+30			4.10		209.65	sewer manhole
12+50			4.20		209.55	
BM 2				4.60	209.15	3 in alum. disk stamped "City Surveyor" at SW corner of Rio Grande and Second
					< 209.15 >	
$\sum$ = 6.75 − 9.70 = −2.95						
BM 1 = 212.10 − 2.95 = 209.15 BM 2 check						

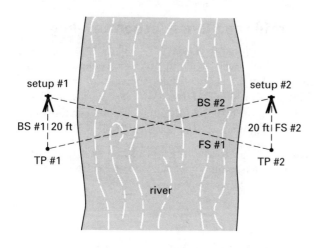

Figure 3.4 *Reciprocal Leveling*

Table 3.2 *Bench Circuit Notes*

setup I	BS	HI	FS	elev
TP I	5.20	x		222.54
TP II			6.14	
setup II				
TP I	4.10	x		222.54
TP II			5.10	
sum	9.30		11.24	
avg.	4.65		5.62	
difference in elevation = 4.65 − 5.62 = −0.97				
El. TP II = 222.54 − 0.97 = 221.57				

Practice Problems

1. What is a point called that is used when the initial setup location is too far away?

   (A) a benchmark

   (B) a backsight

   (C) a turning point

   (D) a intermediate foresight

2. A leveling rod is held on a point whose elevation is to be determined. The distance to the rod is 153 ft, and the FS reading is 5.30. What is the elevation of the point if the HI is 530.00 ft?

   (A) 524.70 ft

   (B) 535.30 ft

   (C) 677.70 ft

   (D) 688.30 ft

3. Benchmarks established as a basis of control for lower-order work are usually established using what procedure?

   (A) precise leveling

   (B) spirit leveling

   (C) trigonometric leveling

   (D) differential leveling

4. You are using an engineer's level with an interval factor $K = 101$ and a stadia constant $C = 1.10$. The point you are occupying has an elevation of 357.55 ft. Given the following reading, what is the elevation of the point you are sighting?

$$HI = 5.25 \text{ ft}$$
$$\text{upper crosshair reading} = 7.20 \text{ ft}$$
$$\text{lower crosshair reading} = 5.00 \text{ ft}$$
$$\text{vertical angle} = -2°25'$$

   (A) 222.90 ft

   (B) 347.29 ft

   (C) 348.14 ft

   (D) 366.11 ft

5. What is the most widely used self-reading rod?

   (A) spirit rod

   (B) Lenker rod

   (C) Chicago rod

   (D) Philadelphia rod

6. What are the notes called that are kept when using a Lenker rod?

   (A) peg notes

   (B) precise notes

   (C) three-wire notes

   (D) no notes need to be taken

7. Given the following leveling circuit notes, what is the adjusted elevation of the temporary benchmark (TBM 1)?

sta	BS (+)	HI	FS (−)	elev	adj elev
levels for Pacific Avenue 2/20/94 C. Landmann–Instrument 72°F clear P. Cuomo–Rod					
BM 1				330.62	(330.62)
	3.15				
TP 1			3.62		
	2.98				
TP 2			3.98		
	2.75				
TBM 1			4.73		
	3.66				
TP 3			5.61		
	4.02				
TP 4			5.99		
	4.42				
TP 5			6.09		
	5.04				
BM 2			6.53		(320.125)

(A) 319.09 ft

(B) 320.38 ft

(C) 327.19 ft

(D) 327.26 ft

8. Given the following leveling circuit notes, what is the elevation of the centerline at sta 22+50?

sta	BS	HI	SS	FS	elev
centerline profile of Aliso Avenue 2/21/94 C. Landmann–Instrument 72°F cloudy P. Cuomo–Rod					
BM 1	4.02				454.69
20+00			5.20		
20+50			5.30		
21+00			5.40		
21+50			5.50		
TP 1	4.20			6.02	
21+80			5.80		
22+00			5.95		
22+50			6.05		
BM 2				5.83	451.06

(A) 448.64 ft

(B) 450.84 ft

(C) 452.66 ft

(D) 457.11 ft

9. If you are establishing a benchmark across a wide river, what is the best procedure to use?

(A) differential leveling

(B) trigonometric leveling

(C) reciprocal leveling

(D) spirit leveling

10. You have discovered that the rod that you were using was 0.04 ft short of the standard length. What should you do to correct your leveling circuit?

(A) Subtract 0.04 ft from all elevations.

(B) Add 0.04 ft to all elevations.

(C) Subtract 0.04 ft from all turning points.

(D) Make no corrections.

Solutions

1. A turning point is used when the initial setup is too far away.

Answer (C)

2. Subtract the FS from the HI to get the elevation point.

$$
\begin{array}{r}
530.00 \text{ ft} \\
-\quad 5.30 \text{ ft} \\
\hline
524.70 \text{ ft}
\end{array}
$$

Answer (A)

3. Precise leveling is spirit leveling of a high order of accuracy, usually extended over large areas, to furnish accurate vertical control as a basis of control for lower order work.

Answer (A)

4.

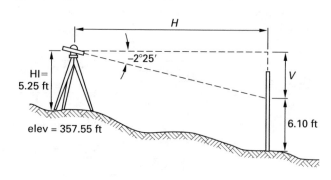

$$ H = (KS \cos \alpha + C)(\cos \alpha) $$

$$ = [(101)(2.20 \text{ ft})(\cos 2°25') + 1.10](\cos 2°25') $$

$$ = (223.10 \text{ ft})(\cos 2°25') $$

$$ = 222.90 \text{ ft} $$

$$ V = (KS \cos \alpha + C)(\sin \alpha) $$

$$ = [(101)(2.20 \text{ ft})(\cos 2°25') + 1.10](\sin 2°25') $$

$$ = (223.10 \text{ ft})(\sin 2°25') $$

$$ = 9.41 \text{ ft} $$

$$ \text{elev at inst} + \text{HI} \pm V - \text{RR} = \text{elev at rod} $$

$$ 357.55 \text{ ft} + 5.25 \text{ ft} - 9.41 \text{ ft} - 6.10 \text{ ft} = 347.29 \text{ ft} $$

Answer (B)

5. The Philadelphia rod is the most widely used self-reading rod.

Answer (D)

6. A Lenker rod is self-reducing. It is not necessary to keep any leveling circuit notes.

Answer (D)

7.

sta	BS (+)	HI	FS (+)	elev	adj elev
BM 1				330.62	(330.62)
	3.15	333.77			
TP 1			3.62	330.15	330.16
	2.98	333.13			
TP 2			3.98	329.15	329.16
	2.75	331.90			
TBM 1			4.73	327.17	**327.19**
	3.66	330.83			
TP 3			5.61	325.22	325.24
	4.02	329.24			
TP 4			5.99	323.25	323.28
	4.42	327.67			
TP 5			6.09	321.58	321.61
	5.04	326.62			
BM 2			6.53	320.09	(320.125)
Σ	26.02		36.55		
diff = 26.02 − 36.55 = −10.53					
BM 2 = BM 1 − 10.53 = 330.62 − 10.53 = 320.09					

The leveling circuit does not close for −0.035 in 7 turns.
Adjust TP 1 + 0.005, TP 2 + 0.01, etc.

Answer (C)

8.

sta	BS	HI	SS	FS	elev
BM 1	4.02	458.71			454.69
20+00			5.20		453.51
20+50			5.30		453.41
21+00			5.40		453.31
21+50			5.50		453.21
TP 1	4.20	456.89		6.02	452.69
21+80			5.80		451.09
22+00			5.95		450.94
22+50			6.05		**450.84**
BM 2				5.83	451.06

Answer (B)

9. When the foresights and backsights cannot be balanced, reciprocal leveling is used.

Answer (C)

10. No corrections would be necessary because you are adding the backsights and subtracting the foresights, thus canceling out any error in rod length.

Answer (D)

Angle Measurement

4

1. Introduction

Angles are one of the primary parameters used by surveyors in describing points in relation to one another. While the actual measurement of angles is performed using a single method (a *transit*), there are alternative ways to describe a measured angle, such as *bearings*, *azimuths*, and *traverses*. Important differences exist between the methods, and computational methods require knowledge of all three.

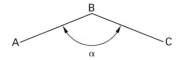

(a) included angle, internal angle, or explement angle

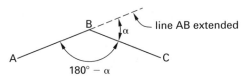

(Use right for clockwise rotation
or left for counterclockwise
from the extension of line AB.)

(b) deflection angle

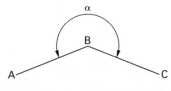

(c) exterior angle or station angle

Figure 4.1 *Types of Angles*

2. Bearings

A. Definition

The *bearing* of a line is the direction of the line with respect to any given meridian and is described by the 90° quadrant (I, II, III, or IV) in which the line falls and by the acute angle between the line and the meridian within the quadrant. The *reference meridian* used is either true, magnetic, or grid, and the bearings will similarly be true, magnetic, or grid, in keeping with the reference meridian. A bearing notation will always state first the initial direction (north or south) and then the

angle (east or west). Thus, bearings will never have a value of more than 90° because this would put the line in a different quadrant.

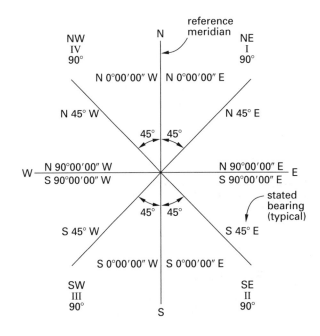

Figure 4.2 *Relationship of Bearings to the Reference Meridian*

B. True Bearings

A *true meridian* is an imaginary line drawn pole-to-pole on the earth's surface at any location desired. Thus, more technically stated, a bearing whose reference meridian is a true meridian passing through the earth's north and south geodetic poles, whose projection passes through the celestial pole (geodetic north), and whose locus is defined by its angular relationship to Polaris, is a *true bearing*. For any given point on earth, the true meridian is always the same; therefore, directions referenced to the true meridian will remain the same regardless of time. A bearing on a map prepared in 1892 shown as N 50°20'18" W and based on true north will be the same bearing today.

Example 4.1

The bearing of line AB is N 50°16'08" E, and the bearing of line AC is N 64°26'32" E. What is angle BAC?

Solution

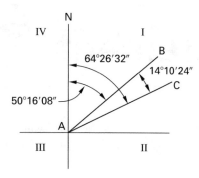

To find the angle between two lines in the same quadrant, subtract the values of the bearings.

$$\text{N } 64°26'32'' \text{ E} - \text{N } 50°16'08'' \text{ E} = 14°10'24''$$

Example 4.2

The bearing of line AB is N 50°16′08″ E, and the bearing of line AC is S 64°26′32″ E. What is angle BAC?

Solution

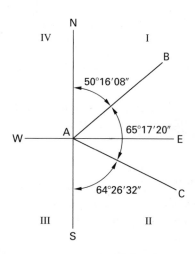

The angle between a line in quadrant I and a line in quadrant II (or a line in quadrant III and a line in quadrant IV) is found by adding the values of the bearings and subtracting the sum from 180°.

$$180° - (50°16'08'' + 64°26'32'') = 65°17'20''$$

Example 4.3

The bearing of line AB is N 50°16′08″ E, and the bearing of line AC is S 64°26′32″ W. What is angle BAC?

Solution 1

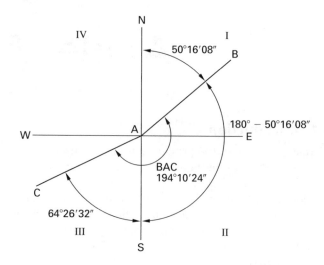

The angle between a line in quadrant I and a line in quadrant III (or a line in quadrant IV and a line in quadrant II) is found as follows.

step 1: Subtract the value of the bearing of the line in quadrant I from 180°.

$$180° - 50°16'08'' = 129°43'52''$$

step 2: Add the solution to the value of the bearing of the line in quadrant III.

$$129°43'52'' + 64°26'32'' = 194°10'24''$$

Solution 2

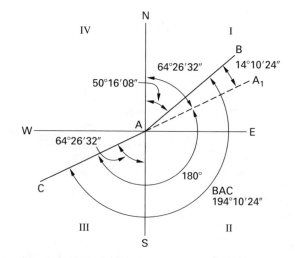

Extend line CA into quadrant I (Line A-A_1). This bearing is N 64°26′32″ E.

(a) Subtract the value of the bearing AB from the value of bearing A-A_1.

$$64°26'32'' - 50°16'08'' = 14°10'24''$$

(b) Add the result to 180°.

$$180° + 14°10'24'' = 194°10'24''$$

Example 4.4

The bearing of line AB is N 50°16′08″ E, and the bearing of line AC is N 64°26′32″ W. What is angle BAC?

Solution

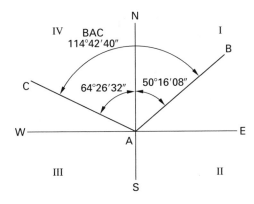

The angle between a line in quadrant I and quadrant IV (or a line in quadrant II and a line in quadrant III) is found by adding the values of the bearings.

$$50°16'08'' + 64°26'32'' = 114°42'40''$$

C. Magnetic Bearings

A *magnetic bearing* is a bearing whose reference meridian is magnetic north as taken by a compass. Since the magnetic poles are at some distance from the true poles, the magnetic meridian is not parallel with the true meridian. The location of the magnetic poles is constantly changing, and the magnetic bearing between points does not remain constant.

The angle between a true meridian and the magnetic meridian at the same point is called the *magnetic declination*. A line on the earth's surface having the identical magnetic declination throughout its length is called an *isogonic* line. A line where the magnetic and true meridians coincide is called an *agonic* line. Throughout California, the magnetic declination is east—that is, the magnetic pole lies east of the true pole. The declination increases as the latitude of the observer increases from south to north. In California, the declination increases from approximately 14° E to approximately 20° E.

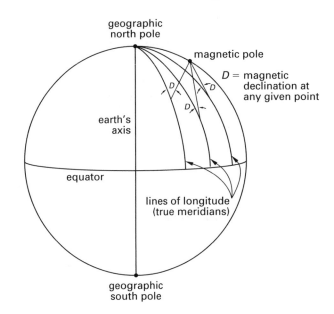

Figure 4.3 *Relationship of Geographic Pole to Magnetic Pole*

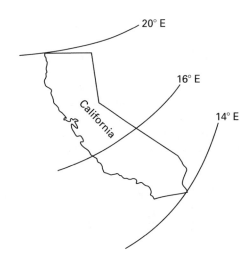

Figure 4.4 *Isogonic Chart (1960)*

Example 4.5

In 1910, the magnetic bearing of line AB was N 36°15′ W, and the magnetic declination was 2°25′ W. The present magnetic declination is 3°45′ E. What are the true and magnetic bearings of the line at present?

Solution

Determine the true bearing of AB by adding the declination to the acute angle between magnetic north from 1910 and line AB.

$$\text{true bearing} = \text{N } 36°15' \text{ W} + 2°25'$$
$$= \text{N } 38°40' \text{ W}$$

Determine the magnetic bearing of line AB by adding $3°45'$ to the magnetic bearing from 1910.

$$\text{magnetic bearing} = \text{N } 38°40' \text{ W} + 3°45'$$
$$= \text{N } 42°25' \text{ W}$$

D. Grid Bearings

A *grid bearing* is derived from the acute angle as measured clockwise or counterclockwise from a reference meridian of the plane coordinate projections system used. Simply stated, an imaginary grid is superimposed onto a map and the angles measured between any ground location and the north-south grid lines is determined.

3. Azimuths

The *azimuth* of a line is its direction, given by the angle between the meridian and the line, measured in a clockwise direction. It ranges from 0° to 360°. As with bearings, azimuths can be either true, magnetic, or grid, depending on the reference meridian. Azimuths can be indicated from either the south point or the north point of a meridian, but they are always measured in a clockwise direction. The U.S. Coast and Geodetic Survey formerly indicated south as the zero direction for its work; they now use north.

A. True Azimuths

A *true azimuth* is an azimuth whose reference meridian is a true meridian passing through the earth's north and south geodetic poles. The projection of a true azimuth passes through the celestial pole (geodetic north). The position of the azimuth is defined by its angular relationship to the star Polaris. For any given point on earth, the true meridian is always the same. Directions referenced to the true meridian will remain the same regardless of time.

B. Magnetic Azimuths

As with the magnetic bearings, a *magnetic azimuth* is the angle between the vertical plane through an observed object and the vertical plane in which a compass needle will come to rest (magnetic north).

C. Grid Azimuths

At the point of observation, the *grid azimuth* is the angle in the plane of the projection measured between the central meridian of the plane coordinate projection system and a line containing the object sighted.

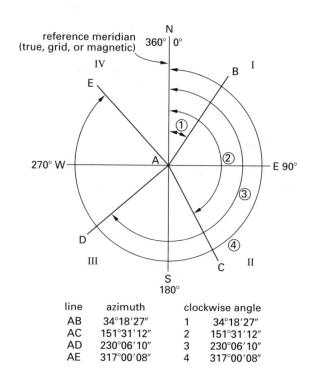

line	azimuth		clockwise angle
AB	34°18'27"	1	34°18'27"
AC	151°31'12"	2	151°31'12"
AD	230°06'10"	3	230°06'10"
AE	317°00'08"	4	317°00'08"

***Figure 4.5** Azimuths*

Example 4.6

The azimuth of line AB is $62°15'37''$, and the azimuth of line AC is $247°30'56''$. What is the horizontal angle BAC?

Solution

To find the horizontal angle between two azimuths, subtract the azimuths.

$$\angle\text{BAC} = 247°30'56'' - 62°15'37''$$
$$= 185°15'19''$$

Example 4.7

The azimuth of line AB is $75°32'08''$.

The azimuth of line AC is $165°03'52''$.

The azimuth of line AD is $227°11'45''$.

The azimuth of line AE is $326°18'26''$.

What are the bearings of lines AB, AC, AD, and AE?

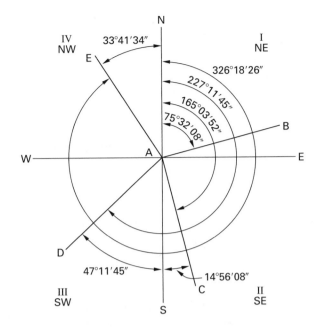

Solution

Line AB is in quadrant I and is $75°32'08''$ east of north, bearing AB $= \text{N } 75°32'08'' \text{ E}$. The bearing of a line in quadrant I is northeast and is the same as the value of the azimuth.

Line AC is in quadrant II. The value of the bearing of line AC is equal to $180°$ minus the value of the azimuth.

$$\text{bearing AC} = 180° - 165°03'52''$$
$$= \text{S } 14°56'08'' \text{ E}$$

Line AD is in quadrant III. The value of the bearing of line AD is equal to the value of the azimuth minus $180°$.

$$\text{bearing AD} = 227°11'45'' - 180°$$
$$= \text{S } 47°11'45'' \text{ W}$$

Line AE is in quadrant IV. The value of the bearing of line AE is equal to $360°$ minus the value of the azimuth.

$$\text{bearing AE} = 360° - 326°18'26''$$
$$= \text{N } 33°41'34'' \text{ W}$$

4. Traverses

Traversing involves relating points through angles and distances. The two types of traversing are the open traverse and closed traverse.

The field procedure for performing an *open traverse* is to set the instrument on station B, sight station A, and then measure the angle and distance to station 1. Here, the angle measured to station A is called the *backsight*

(BS), and the angle measured to station 1 the *foresight* (FS). Then the instrument is moved to station 1. A backsight reading is taken on station B and again on station 2, which is the new foresight. This procedure is repeated until the last angle at station C is measured. The bearings of the traverse line are computed by using the measured angles, and then the measured bearing of line CD is compared to the known bearing. If the measured bearing of line CD does not match the known bearing but falls within error guidelines, the known bearing is held fixed and the bearings of the traverse line are adjusted (corrected) accordingly to either the compass rule or the transit rule.

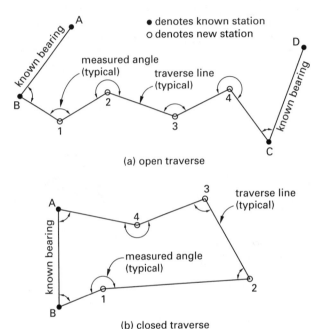

Figure 4.6 Traverses

The procedure for performing a *closed traverse* is basically the same as for performing the open traverse, except the survey starts and ends on line AB, with point A termed the *point of beginning*. When the survey is complete, the angles are adjusted (corrected) according to either the compass rule or the transit rule. In the closed case shown in Fig. 4.6, the sum of the interior angles must add up to $(n-2)(180°)$, where n is the number of legs.

$$\text{sum of interior angles} = (n-2)(180°)$$
$$= (6-2)(180°)$$
$$= (4)(180°)$$
$$= 720°$$

Prior to computing bearings for each line of a closed traverse, the measured angles must be adjusted to add up to $720°$.

Example 4.8

An open traverse is performed in the field. The starting station is station Big, and the bearing of the line between Big and station Little is N 50°07′52″ W. This line is the *basis for bearings* for the survey. The survey will close on station Red, and the line between Red and station Blue has a recorded bearing of S 80°57′08″ E. The results of the survey are as follows.

instrument station	BS	FS	angle
Big	Little	sta 1	69°07′12″ rt
sta 1	Big	sta 2	251°20′31″ rt
sta 2	sta 1	Red	111°32′15″ rt
Red	sta 2	Blue	257°10′50″ rt

What is the measured bearing of the line Red-Blue?

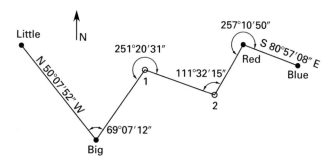

Solution

Using the measured angles, the bearings of the lines of the traverse are computed starting with line Big-sta 1.

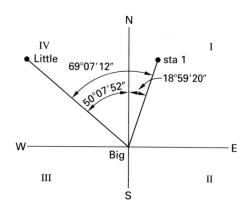

The lines in question fall in quadrants I and IV; therefore, the bearing of the line Big-sta 1 is

$$69°07′12″ - 50°07′52″ = \text{N } 18°59′20″ \text{ E}$$

For line sta 1-sta 2, see the following illustration.

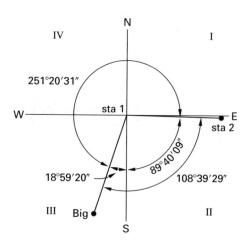

The lines in question fall in quadrants II and III. The computation will be simplified by applying the value of the interior angle.

$$360° - 251°20′31″ = 108°39′29″$$

Subtracting the value of the bearing of line sta 1–Big,

$$108°39′29″ - 18°59′20″ = 89°40′09″$$

Since this value is less than 90°, the bearing is in quadrant II and is S 89°40′09″ E.

For line sta 2-Red, see the following illustration.

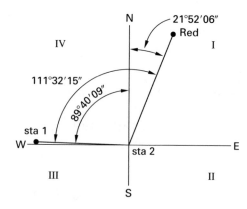

The lines in question fall in quadrants I and IV. Therefore, the bearing of line sta 2-Red is the measured angle minus the value of the bearing between sta 1 and sta 2.

$$111°32′15″ - 89°40′09″ = 21°52′06″$$
$$= \text{N } 21°52′06″ \text{ E}$$

For line Red-Blue, see the following illustration.

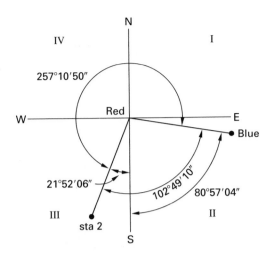

The lines in question are in quadrants II and III. To simplify the computation, subtract the measured angle from 360°. This will give the value of the interior angle.

$$360° - 257°10'50'' = 102°49'10''$$

Since the lines are in quadrants III and IV, subtract the value of the bearing of line Red-sta 2 from 102°49'10'' to compute the bearing of line Red-Blue.

$$102°49'10'' - 21°52'06'' = 80°57'04''$$
$$= S\ 80°57'04''\ E$$

The difference between the record and measured bearing of Red-Blue is only 04''. This is well within tolerance.

Practice Problems

1. What are the directions of bearings in quadrant I?
   (A) NE
   (B) SE
   (C) SW
   (D) NW

2. What are the directions of bearings in quadrant III?
   (A) NE
   (B) SE
   (C) SW
   (D) NW

3. A line that bears N 58°40' E also bears which of the following?
   (A) N 31°20'' W
   (B) S 58°40' W
   (C) S 58°40' E
   (D) none of the above

4. Which of the following is true for a line that bears S 37°12'18'' W?
   (A) It has an azimuth of 217°12'18''.
   (B) It makes a clockwise angle of 37°12'18'' with the meridian, starting at south.
   (C) It has a reverse direction of N 37°12'18'' E.
   (D) All of the above are true.

5. Line AB bears N 38°16'27'' W. Line AC bears N 42°11'36'' E. What is angle BAC?
   (A) 03°55'09''
   (B) 80°28'03''
   (C) 99°31'57''
   (D) none of the above

6. Line AB bears N 50°12'42'' E. Line AC bears S 11°19'16'' E. What is angle BAC?
   (A) 38°53'26''
   (B) 61°31'58''
   (C) 118°28'02''
   (D) 298°28'02''

7. In 1910, the magnetic declination at a point was 05°12' E. In 1993 it was 11°20' E. If line AB had a magnetic bearing of N 52°30' E in 1910, what was the magnetic bearing of the line in 1993?
   (A) N 46°22' E
   (B) N 63°50' E
   (C) N 57°42' E
   (D) N 69°02' E

8. Using the data in Prob. 7, what is the true bearing of line AB?
   (A) N 57°42' E
   (B) N 46°22' E
   (C) N 41°10' E
   (D) N 63°50' E

9. Convert each of the following bearings to azimuths.

   (A) N 67°30′12″ E

   (B) N 11°42′30″ W

   (C) S 21°09′54″ E

   (D) S 70°16′39″ W

10. Convert each of the following azimuths to bearings.

   (A) 336°11′32″

   (B) 98°08′08″

   (C) 215°37′18″

   (D) 88°01′57″

11. Line AB bears S 17°19′37″ W. Angle BAC measured counterclockwise is 65°29′53″. What is the bearing of line AC?

   (A) S 82°49′30″ E

   (B) S 48°10′16″ W

   (C) S 48°10′16″ E

   (D) S 82°49′30″ W

12. Station Larry is occupied by the instrument, and two sets of direct and reverse readings are made between stations Curly and Moe. The line Larry-Curly bears N 50°10′08″ W. Based on the following set of notes, what is the bearing of line Larry-Moe?

	inst sta			direct	reverse
set 1	Larry	BS = Curly		00°00′10″	180°00′14″
		FS = Moe		155°28′36″	335°28′44″
set 2	Larry	BS = Curly		20°00′06″	200°00′10″
		FS = Moe		175°28′46″	355°28′38″

Solutions

1. Answer (A)

2. Answer (C)

3. The reverse bearing is 180° from forward bearing.

   Answer (B)

4. Answer (D)

5. The lines are in quadrants IV and I. Add the values of the bearings to compute the angle.

$$38°16′27″ + 42°11′36″ = 80°28′03″$$

   Answer (B)

6. The lines are in quadrants I and II. Add the values of the bearings and subtract the sum from 180°.

$$180° - (50°12′42″ + 11°19′16″) = 118°28′02″$$

   Answer (C)

7. The angle between magnetic north in 1910 and in 1993 equals $11°20′ - 05°12′ = 06°08′$. The shift is to the east. The magnetic bearing in 1993 was N 52°30′ E − 06°08′ = N 46°22′ E.

   Answer (A)

8. Declination is east of the meridian. The true bearing equals the magnetic bearing plus the declination.

$$N 52°30′ E + 05°12′ = N 57°42′ E$$

   Answer (A)

9. (A) Azimuths in quadrant I equal the bearing.
       az = 67°30′12″

   (B) Azimuths in quadrant IV equal 360° minus the value of the bearing.
       az = 360° − 11°42′30″ = 348°17′30″

   (C) Azimuths in quadrant II equal 180° minus the value of the bearing.
       az = 180° − 21°09′54″ = 158°50′06″

   (D) Azimuths in quadrant III equal 180° plus the value of the bearing.
       az = 180° + 70°16′39″ = 250°16′39″

10. (A) 360° − 336°11′32″ = N 23°48′28″ W

   (B) 180° − 98°08′09″ = S 81°51′52″ E

   (C) 215°37′18″ − 180° = S 35°37′18″ W

   (D) 88°01′57″ = N 88°01′57″ E

11. Line AB is in quadrant III. Line AC is in quadrant II. Bearing AC equals

$$BAC - \text{value of bearing } AB = 65°29′53″ - 17°19′37″$$
$$= S 48°10′16″ E$$

   Answer (C)

12. Set 1: BS − avg seconds D & R = 12″ or
$$(10″ + 14″)/2$$
 FS − avg seconds D & R = 40″ or
$$(36″ + 44″)/2$$

Set 2: BS − avg seconds D & R = 08″ or
$$(06″ + 10″)/2$$
 FS − avg seconds D & R = 42″ or
$$(46″ + 38″)/2$$

Direction: degrees and minutes = 155°28′ from foresight to point Moe

Average seconds:
 position 1: 40″ − 12″ = 28″
 position 2: 42″ − 08″ = 34″
 avg = 31″ or
$$(28″ + 34″)/2$$

Angle Curly-Larry-Moe equals 155°28′31″, or 155°28′ plus 31″.

Find the bearing Larry-Moe:

step 1: Convert bearing Larry-Curly.

azimuth − 360° − 50°10′08″ = 309°49′52″

step 2: Add angle Curly-Larry-Moe to azimuth Larry-Curly.

309°49′52″ + 155°28′31″ = 465°49′23″

step 3: Subtract 360° from the results of step 2.

465°49′23″ − 3⟨…⟩18′23″

⟨…⟩-Moe

step 4: Convert azi⟨…⟩

bearing − 180° − 105°18′23″ = ⟨…⟩

Larry-Moe is in quadrant II, therefore the bear⟨…⟩ is S 74°41′37″ E.

Error Analysis and Distribution

Contributed by Jeremy Evans, PLS

1. Introduction

Surveying involves two separate yet related ways of describing and treating errors in measurements of any kind: the *level of accuracy* and the *distribution of error (balancing)* assigned to each measurement.

The various methods of distributing the error generated in performing a traverse are discussed in Ch. 6. The most common method of distributing the error generated in performing a leveling circuit is to simply prorate an equal value of error to each elevation measured, as discussed in Ch. 3. The *error* distributed by any of these methods can be the result of any shortcomings of the procedure or method of measurement, the degree of care exercised by the survey party, the calibration frequency of the equipment used, and the weather and terrain encountered. Generally, the numerical value of error to be distributed is determined by comparing the actual field readings with those computed.

The *level of accuracy*, however, is a measure of uncertainty regarding the survey overall and can be shown mathematically as a *tolerance*. This chapter covers the basics of the statistical theory used to develop the level of accuracy for a survey taken as a whole.

2. Data and Error Analysis

All measurements have two components: for example, $100.0' \pm 0.02'$. The first component is the measurement itself. The second component is the statement of uncertainty, which is based on statistical analysis of the equipment and procedures used. If no statement of uncertainty is given, then it is assumed to be one-half of the least count of the measurement. For example, 100.00 ft is assumed to have an uncertainty of $\pm 0.005'$. An angle of $90°01'15''$ is assumed to have an uncertainty of $\pm 0.5''$. *Random error theory* is used to determine the value of uncertainty in a set of measurements.

Suppose an angle is repeatedly measured and noted as in Table 5.1.

The differences in each of the angles shows the inability of the person making the measurements to make perfect observations. Assuming that all systematic errors have been removed, the results can be analyzed to determine just how precisely this measurement has been made.

Table 5.1 Angle Measurements

#	measurement
1	$100°15'36''$
2	$100°15'42''$
3	$100°15'30''$
4	$100°15'33''$
5	$100°15'40''$
6	$100°15'35''$
7	$100°15'37''$
8	$100°15'36''$
9	$100°15'31''$
10	$100°15'43''$
11	$100°15'40''$
12	$100°15'33''$
13	$100°15'35''$
14	$100°15'37''$
15	$100°15'36''$
16	$100°15'35''$
total	$579''$

First, the mean is calculated by summing the measurements and dividing by the number of measurements, n.

The mean for the measurements in Table 5.1 is

$$100°15'00'' + (579''/16) = 100°15'36''$$

The residual for each individual measurement is then determined. The *residual*, r, is the individual-measurement mean, taken as the difference between the first measurement and each successive measurement. For the first measurement, the residual is

$$r = 36'' - 36'' = 0 \quad \begin{bmatrix} \text{using only the} \\ \text{necessary digits} \end{bmatrix}$$

For the second measurement,

$$r = 42'' - 36'' = 6''$$

For the third,

$$r = 30'' - 36'' = -6''$$

The residuals are determined for each measurement in Table 5.2.

Table 5.2 Angles and Residuals

#	measurement	r (inches)	r^2 ((inches)^2)
1	100°15′36″	0	0
2	100°15′42″	6	36
3	100°15′30″	−6	36
4	100°15′33″	−3	9
5	100°15′40″	4	16
6	100°15′35″	−1	1
7	100°15′37″	1	1
8	100°15′36″	0	0
9	100°15′31″	−5	25
10	100°15′43″	7	49
11	100°15′40″	4	16
12	100°15′33″	−3	9
13	100°15′35″	−1	1
14	100°15′37″	1	1
15	100°15′36″	0	0
16	100°15′35″	−1	1
total	579″		201

Occasionally, the standard deviation is called the *mean squared error* or the *standard error*. The standard deviation establishes the limits within which measurements are expected to fall 68.27% of the time. It is an uncertainty statement regarding the individual measurements compared to the mean, and therefore is an indication of the level of accuracy of the measurements. The formula for determining the standard deviation is

$$\sigma = \pm\sqrt{\frac{\Sigma r^2}{n-1}} \qquad 5.1$$

The sum of the residuals squared is divided by the number of measurements. In Table 5.2, $\Sigma r^2 = 201$. Since $n = 16$,

$$\sigma = \pm\sqrt{\frac{201 \text{ in}^2}{16-1}} = \pm 3.7''$$

In the example, 68.27% (or one standard deviation) of the individual measurements should fall between $100°15'36'' - 3.7'' = 100°15'32.3''$ and $100°15'36'' + 3.7'' = 100°15'39.7''$. Checking the residuals shows that 10 (or 63%) of the measurements fall in the standard deviation range rather than 11 as expected. The difference here between 63% and 68.27% is due to the relatively few individual measurements made. The more measurements taken, the closer to 68.27% of the measurements would fall within one standard deviation of the mean.

3. Standard Error of the Mean

The *standard error of the mean*, sometimes referred to as the *root mean squared error*, is an uncertainty statement regarding the mean value and not the individual measurements that make up the mean (as was the case with the standard deviation). In this case, the standard error of the mean is the mean value versus the theoretical value and is an indication of the accuracy of the measurements. The formula to determine the standard error of the mean is

$$\sigma_m = \pm\frac{\sigma}{\sqrt{n}} \qquad 5.2$$

For this example,

$$\sigma_m = \pm\frac{3.7''}{\sqrt{16}}$$
$$= \pm 0.92'' \quad (\pm 1'')$$

This result indicates that the mean in the example is within the range of $\pm 1''$ from the theoretical value.

4. Groups of Measurements

Suppose a line is measured in several parts and the standard deviation is determined for each part as previously described. To determine the standard deviation for the entire line (since the error for each part is different), use the formula for the error in a sum. σ_{sum} is the standard deviation of the sum of the parts, and σ_1, σ_2, and so on are the individual standard deviations.

$$\sigma_{sum} = \pm\sqrt{\sigma_1^2 + \sigma_2^2 + \sigma_3^2 + \cdots + \sigma_n^2} \qquad 5.3$$

5. Classifications of Accuracy

When surveys are done for control, boundary, or construction purposes and certain accuracies are required, the survey will probably need to be performed to a specified *classification of accuracy*. The primary reason for the detailed specifications and procedures is to ensure that the required accuracy is attained or exceeded throughout the entire survey. A survey of a given classification must conform to all requirements for that classification.

Table 5.3 is a portion of Table 4-04-B, Classifications of Accuracy, of the *Caltrans Survey Manual*. Two classifications are given: second order (modified) and third order. Second order (modified) is of greater precision than third order. The higher the order of accuracy required, the longer the survey will take, and therefore, the costlier it will be.

Table 5.3 *Classification of Accuracy Standards*

Traverse

	second order (modified)	third order
maximum number of courses between checks for azimuth	15	25
azimuth closure not to exceed*	10″ 3.0″ per station	30″ 8.0″ per station
position closure (after azimuth adjustment) not to exceed*	1.67′ 1:10,000	3.34′ 1:5,000
distance measurement accurate within	1:15,000	1:7,500
minimum distance to be measured with EDMs	0.1 mile	0.05 mile
minimum number of angle observations: a. one-second theodolite b. one-minute or 20-second theodolite	4 pos. 1 set of 6D, 6R	2 pos. 1 set of 2D, 2R

*The expressions for closing errors in traverse surveys are given in two forms. The formula that gives the smaller permissible closure should be used.

Reprinted by permission of CALTRANS.

Other reasons for conforming to specified classifications of accuracy are to establish uniformity among different surveys, to minimize oversurveying, and to provide requirements beyond the familiar linear-closure standards.

Standards and procedures also can prevent or minimize oversurveying and thus keep costs to a minimum. Under most conditions, the procedures given in the Classification of Accuracy Standards provide closure and relative position accuracies well within the standards specified. Such standards tend to be very conservative, thus allowing even severe site conditions to fall within the accuracies required.

Practice Problems

1. What are the two error types?

2. Explain the difference between accuracy and precision as they apply to surveying.

3. Calculate (A) the most probable value (mean), (B) the standard deviation of a single measurement, and (C) the standard error of the mean from the following measurements of the same angle.

48°32′00″ 48°31′50″ 48°32′20″ 48°31′40″ 48°32′10″
48°31′55″ 48°32′07″ 48°32′02″ 48°31′52″ 48°32′15″
48°31′56″ 48°31′45″ 48°32′12″ 48°31′49″ 48°32′04″
48°32′00″

4. A line was measured in six segments with the following results.

segment	distance	σ_m
AB	222.67′	±0.12′
BC	413.66′	±0.17′
CD	116.55′	±0.08′
DE	616.99′	±0.14′
EF	333.09′	±0.05′
FG	815.26′	±0.19′

(A) What is the length of line AG?

(B) What is the standard error of the mean for the entire line?

5. A party chief estimates that both he and his chainman can read and mark each tape end to ±0.01′ using a 100 ft tape. What is the uncertainty in a 1000 ft distance laid out with this tape due to this error source?

6. Three distance segments, each having an uncertainty of ±0.05′, were added to four other segments, each having an uncertainty of ±0.10′. What is the uncertainty in the sum of these seven segments?

7. A line AB has an uncertainty of $\pm0.10'$ when measured once. How many times must this line be measured to reduce the uncertainty to $\pm0.03'$?

8. Using Table 5.3, what is the maximum positional closure for a 4.5 mi traverse (third order)?

Solutions

1. The two error types are systematic and random.

2. *Accuracy* is the agreement of the mean value and the true value. *Precision* is the agreement among readings of the same observed value.

3.

#	measurement	in seconds	r (inches)	r^2 (inches)
1	48°32'00"	60	0	0
2	48°31'50"	50	−10	100
3	48°32'20"	80	20	400
4	48°31'40"	40	−20	400
5	48°32'10"	70	10	100
6	48°31'55"	55	−5	25
7	48°32'07"	67	7	49
8	48°32'02"	62	2	4
9	48°31'52"	52	−8	64
10	48°32'15"	75	15	225
11	48°31'56"	56	−4	16
12	48°31'45"	45	−15	225
13	48°32'12"	72	12	144
14	48°31'49"	49	−11	121
15	48°32'04"	64	4	16
16	48°32'00"	60	0	0
total		957		1889

(A) $\quad \alpha = \dfrac{957''}{16} = 59.8'' = 60''$

$\quad\quad = 48°32'00''$

(B) $\quad \sigma = \pm\sqrt{\dfrac{\Sigma r^2}{n-1}} = \pm\sqrt{\dfrac{1889''}{15}}$

$\quad\quad = \pm11.2'' \quad (\pm11'')$

(C) $\quad \sigma_m = \pm\dfrac{\sigma}{\sqrt{n}} = \pm\dfrac{11.2''}{\sqrt{16}}$

$\quad\quad = \pm2.8'' \quad (\pm3'')$

4. (A) The length of line AG is 2518.22 ft.

(B) $\sigma_m = \sqrt{\begin{aligned}&(0.12')^2 + (0.17')^2 + (0.08')^2 \\ &+ (0.14')^2 + (0.05')^2 + (0.19')^2\end{aligned}}$

$\quad = \pm0.33 \text{ ft}$

5. The error for each tape length is

$\quad \sigma_{\text{series}} = \pm0.01'\sqrt{2} \quad [0.01' \text{ at each tape end}]$

$\quad\quad = \pm0.014'$

The error for the 1000 ft distance is

$\quad\quad \sigma_{\text{series}} = \pm0.04'$

6. For three segments,

$\quad\quad \sigma_{\text{series}} = \pm0.087'$

For four segments,

$\quad\quad \sigma_{\text{series}} = \pm0.20'$

For all seven segments,

$\quad \sigma_{\text{sum}} = \pm\sqrt{(0.087')^2 + (0.020')^2}$

$\quad\quad = \pm0.218'' \quad (\pm0.22')$

7. $\quad\quad \sigma_m = \pm\dfrac{\sigma}{\sqrt{n}}$

$\quad\quad 0.03' = \pm\dfrac{0.10'}{\sqrt{n}}$

$\quad\quad \sqrt{n} = \dfrac{0.10'}{0.03'}$

$\quad\quad n = \left(\dfrac{0.10'}{0.03'}\right)^2$

$\quad\quad = 11.11$

Therefore, this line must be measured 12 times.

8. $\quad \dfrac{1}{5000} = \dfrac{x}{4.5 \text{ mi}} = 0.0009 \text{ mi} = 4.75'$

The maximum position closure is 4.75'.

Traverses 6

1. Introduction

A *traverse* involves relating several points on the ground to each other in terms of horizontal angles (bearings) and horizontal distances (headings); vertical angles and vertical distances are not involved.

To describe traverses, a surveyor chooses one of the following methods: courses (using either bearings and headings, or azimuths) of each leg, latitudes and departures (which are similar to an x-y-coordinate system) of each point, or rectangular coordinates (x-y-coordinates) of each point. Occasionally, a traverse is described by more than one of these methods. Familiarity with all three methods is required for the special surveying examination.

2. Courses

Courses of a traverse (also termed *legs* of a traverse) can be described either in terms of angles and distances or in terms of azimuths.

A. Bearings and Headings

Bearings (angles) are measured in the field using a transit. The angles are turned from a north-south meridian from one point to another, and they are read showing the relative quadrant, as shown in Fig. 6.1. The *heading* is the horizontal distance in feet between two points.

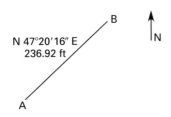

Figure 6.1 *Course*

B. Azimuths

An *azimuth* is measured in the same manner but is expressed in degrees turned clockwise from either the north or south branch of a meridian line. The azimuth for the course shown in Fig. 6.1 is azimuth from the north $47°20'16''$ or azimuth from the south $227°20'16''$. Usually, the azimuth from the north is used.

3. Departures and Latitudes

Simply stated, the change in rectangular coordinates between one point and another in the x-direction is the departure, and the change in the y-direction is the latitude. That is, the *departure* of a line is the distance that the line extends in an east (positive) or west (negative) direction. The *latitude* is the distance that the line extends in a north (positive) or south (negative) direction.

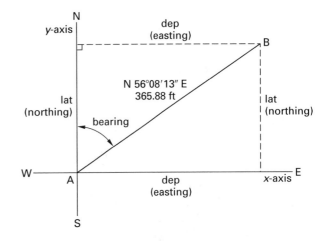

Figure 6.2 *Departures and Latitudes*

The course in Fig. 6.2 is N $56°08'13''$ E, with a heading of 365.88 ft. The latitude of the survey line is the distance north or south along the y-axis from point A to point B. The departure of the survey line is the distance east or west along the x-axis from point A to point B. Latitudes and departures are computed as follows.

$$\text{lat} = \cos\,(\text{bearing or azimuth}) \times (\text{length}) \quad 6.1$$
$$\text{dep} = \sin\,(\text{bearing or azimuth}) \times (\text{length}) \quad 6.2$$

The latitude of line AB in Fig. 6.2 is

$$\text{lat} = \cos\,(56°08'13'')(365.88 \text{ ft}) = 203.87 \text{ ft}$$

The departure of line AB in Fig. 6.2 is

$$\text{dep} = \sin\,(56°08'13'')(365.88 \text{ ft}) = 303.82 \text{ ft}$$

When using latitudes and departures, the algebraic sign of each latitude and departure is important.

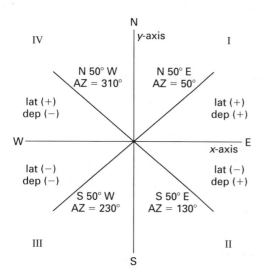

Figure 6.3 *Signs*

To determine the overall difference in latitudes and departures from one end of a traverse to the other, the individual latitudes and departures are summed. Fig. 6.3 shows the algebraic sign to be applied to latitudes and departures depending on which quadrant the line is in. Table 6.1 gives the rules to be applied.

Table 6.1 *Quadrant Definitions*

bearing	quadrant	lat	dep
NE	I	(+)	(+)
SE	II	(−)	(+)
SW	III	(−)	(−)
NW	IV	(+)	(−)

Example 6.1

Compute the latitude and departure for each of the following courses.

(A) N 25°13′07″ E, 135.40 ft

(B) S 30°57′12″ E, 157.90 ft

(C) S 45°18′11″ W, 187.10 ft

(D) N 2°20′26″ W, 144.64 ft

Solution

Applying Eqs. 6.1 and 6.2,

(A) lat = cos (25°13′07″)(135.40 ft) = +122.495 ft
     dep = sin (25°13′07″)(135.40 ft) = +57.690 ft

(B) lat = cos (30°57′12″)(157.90 ft) = −135.413 ft
     dep = sin (30°57′12″)(157.90 ft) = +81.214 ft

(C) lat = cos (45°18′11″)(187.10 ft) = −131.598 ft
     dep = sin (45°18′11″)(187.10 ft) = −132.998 ft

(D) lat = cos (2°20′26″)(144.64 ft) = +144.519 ft
     dep = sin (2°20′26″)(144.64 ft) = −5.907 ft

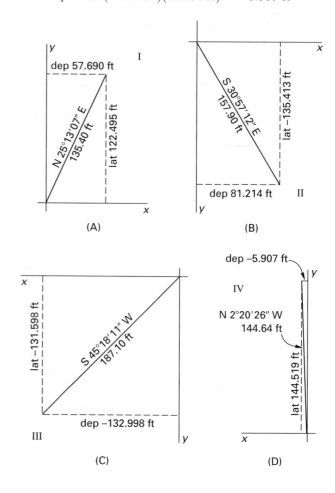

If the course directions are in azimuths instead of bearings, the sign of the cosine or sine of the azimuth is given when hand-held calculators are used. The algebraic sign of the latitude or departure will automatically be assigned correctly.

Example 6.2

A course has an azimuth from the south of 225°18′11″. The length is 187.10 ft. Compute the latitude and departure.

Solution

Applying Eqs. 6.1 and 6.2,

$$\text{lat} = (\cos 225°18'11'')(187.10 \text{ ft})$$
$$= (-0.7034)(187.10 \text{ ft}) = -131.598 \text{ ft}$$

$$\text{dep} = (\sin 225°18'11'')(187.10 \text{ ft})$$
$$= (-0.7108)(187.10 \text{ ft}) = -132.998 \text{ ft}$$

If the latitude and departure of a course are known, the direction and length of the line can be computed by using Eqs. 6.3 through 6.5. This process is known as *inversing* and is shown as follows.

$$\tan(\text{bearing}) = \frac{\text{dep}}{\text{lat}} \qquad 6.3$$

$$\text{length} = \frac{\text{dep}}{\sin(\text{bearing})} \qquad 6.4$$

$$\text{length} = \frac{\text{lat}}{\cos(\text{bearing})} \qquad 6.5$$

Example 6.3

Compute the bearing and distance of the following courses with the following latitudes and departures.

(A) lat: +189.63 ft   dep: −201.78 ft
(B) lat: −300.75 ft   dep: −608.13 ft
(C) lat: −162.18 ft   dep: +211.58 ft
(D) lat: +1100.14 ft   dep: +637.92 ft

Solution

Apply Eqs. 6.3 through 6.5.

(A)
$$\tan = \frac{-201.78 \text{ ft}}{189.63 \text{ ft}} = -1.064072$$
$$\arctan = 46°46'41''$$

From Table 6.1, the line is in quadrant IV.

$$\text{bearing} = \text{N } 46°46'41'' \text{ W}$$
$$\text{length} = \frac{201.78 \text{ ft}}{\sin 46°46'41''} = 276.90 \text{ ft}$$

Check.

$$\text{length} = \frac{\text{lat}}{\cos(\text{bearing})} = \frac{189.63 \text{ ft}}{\cos 46°46'41''}$$
$$= 276.90 \text{ ft}$$

(B)
$$\tan = \frac{-608.13 \text{ ft}}{-300.75 \text{ ft}} = 2.022045$$
$$\arctan = 63°41'07''$$

From Table 6.1, the line is in quadrant III.

$$\text{bearing} = \text{S } 63°41'07'' \text{ W}$$
$$\text{length} = \frac{608.13 \text{ ft}}{\sin 63°41'07''} = 678.43 \text{ ft}$$

Check.

$$\text{length} = \frac{300.75 \text{ ft}}{\cos 63°41'07''} = 678.43 \text{ ft}$$

(C)
$$\tan = \frac{211.58 \text{ ft}}{-162.18 \text{ ft}} = 1.304600$$
$$\arctan = 52°31'45''$$

From Table 6.1, the line is in quadrant II.

$$\text{bearing} = \text{S } 52°31'45'' \text{ E}$$
$$\text{length} = \frac{211.58 \text{ ft}}{\sin 52°31'45''} = 266.59 \text{ ft}$$

Check.

$$\text{length} = \frac{162.18 \text{ ft}}{\cos 52°31'45''} = 266.59 \text{ ft}$$

(D)
$$\tan = \frac{637.92 \text{ ft}}{1100.14 \text{ ft}} = 0.579853$$
$$\arctan = 30°06'27''$$

From Table 6.1, the line is in quadrant I.

$$\text{bearing} = \text{N } 30°06'27'' \text{ E}$$
$$\text{length} = \frac{637.92 \text{ ft}}{\sin 30°06'27''} = 1271.71 \text{ ft}$$

Check.

$$\text{length} = \frac{1100.14 \text{ ft}}{\cos 30°06'27''} = 1271.71 \text{ ft}$$

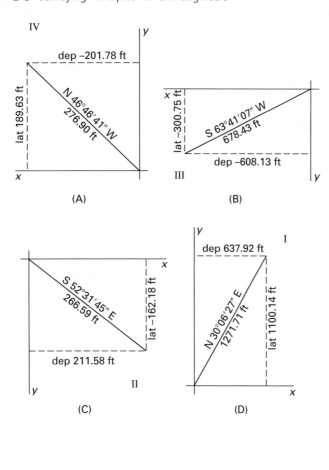

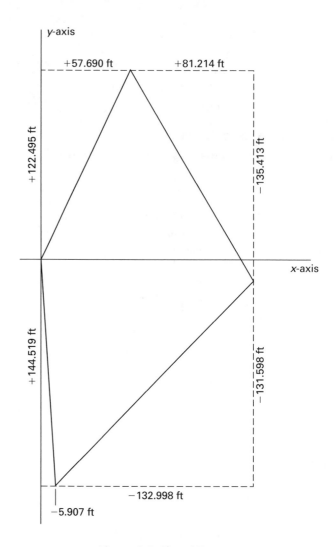

4. Types of Traverses

A. Closed Traverse

A *closed traverse* is several courses that return to the starting point (termed *point of beginning*). An example is shown in Fig. 6.4. The courses in this traverse are the courses used in Ex. 6.1.

In a perfectly closed traverse, the algebraic sum of the latitudes and the algebraic sum of the departures each equal zero. The latitudes and departures from Ex. 6.1 are used in Table 6.2 to illustrate this.

Figure 6.4 Closed Traverse

The algebraic sum of the latitudes is 0.003 ft and of the departures is −0.001 ft. This is caused by rounding in computing the latitude or departure. This amount is insignificant, and for all practical purposes, the traverse closes and is considered balanced. This rarely occurs in the field due to the systematic and random error involved in taking measurements.

Table 6.2 Traverse Data

course	bearing	length	lat +(N)	lat −(S)	dep +(E)	dep −(W)
A	N 25°13′07″ E	135.40	122.495		57.690	
B	S 30°57′12″ E	157.90		135.413	81.214	
C	S 45°18′11″ W	187.10		131.598		132.998
D	N 02°20′26″ W	144.64	144.519			5.907
			267.014	267.011	138.904	138.905

B. Balancing a Closed Traverse

No matter how carefully any survey is performed, not all of the sources of error can be eliminated. When all the courses are plotted based on unadjusted field measurements, the end point of the survey (latitudes and departures, or x-y coordinates) will not be the same as the point of beginning. The survey data must be adjusted to make the end point close on the beginning point. The steps to be followed to make the adjustment are:

step 1: Adjust the field angles such that the sum of the interior angles equals $(n - 2)(180°)$.

step 2: Compute the bearings or azimuths of the courses using the adjusted angles.

step 3: Compute the latitudes and departures of all the courses using Eqs. 6.1 and 6.2.

step 4: Compute the difference between the sum of the north and south latitudes and, separately, the east and west departures.

step 5: Using these differences, compute the total error of closure and the relative error of closure.

step 6: If the relative error of closure is within accepted tolerance, adjust the latitudes and departures such that the traverse closes. If it does not, then more field work will be required.

step 7: Recompute the bearings and distances of the courses using the adjusted latitudes and departures.

Example 6.4

The following illustration (at the top of the next column) shows the results of a field survey of a closed traverse. Check the traverse for closure, and balance it if it does not close.

Solution

step 1: Adjust the field angles by first summing the interior angles of the following field measurements.

$$85°49'42''$$
$$119°21'45''$$
$$84°11'50''$$
$$70°36'31''$$

The sum is $359°59'48''$. It should be

$$(n - 2)(180°) = 360°00'00''$$

Therefore, the sum of the field angles is $12''$ too small.

Then, divide $12''$ by the number of angles.

$$\frac{12''}{4} = 03''$$

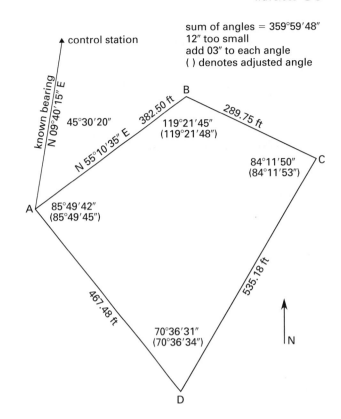

Adding $03''$ to each angle gives the following adjusted field angles.

$$85°49'45''$$
$$119°21'48''$$
$$84°11'53''$$
$$70°36'34''$$

Check.

$$\text{sum} = 360°00'00''$$

step 2: Compute the bearings of the courses using the adjusted angles.

	bearing	length
A-B	N 55°10'35'' E	382.50 ft
B-C	S 64°11'13'' E	289.75 ft
C-D	S 31°36'54'' W	535.18 ft
D-A	N 38°59'40'' W	467.48 ft

step 3: Compute the latitudes and departures of the courses using Eqs. 6.1 and 6.2.

course	lat		dep	
	+(N)	−(S)	+(E)	−(W)
AB	218.4274		313.9996	
BC		126.1676	260.8386	
CD		455.7538		280.5461
DA	363.3287			294.1595
sum	581.7561	581.9214	574.8382	574.7056

step 4: Compute the difference between the sums of the north and south latitudes and the east and west departures.

$$
\begin{aligned}
\text{sum of north latitudes} &= 581.7561 \text{ ft} \\
\text{sum of south latitudes} &= 581.9214 \text{ ft} \\
\text{difference} &= 0.1653 \text{ ft} \\
\text{sum of east departures} &= 574.8382 \text{ ft} \\
\text{sum of west departures} &= 574.7056 \text{ ft} \\
\text{difference} &= 0.1326 \text{ ft}
\end{aligned}
$$

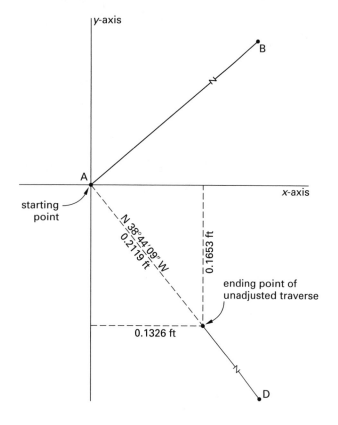

The traverse does not close by 0.1653 ft in the latitudes (northings) and 0.1326 ft in the departures (eastings) as shown in the illustration.

step 5: Compute the error of closure.

The total error of closure is represented by the course from the closing point to the starting point. The bearing (or azimuth) and length of this line is the total error of closure. Use Eqs. 6.3 through 6.5 to compute this course.

$$
\tan(\text{bearing}) = \frac{\text{dep}}{\text{lat}} = \frac{0.1326 \text{ ft}}{0.1653 \text{ ft}}
$$

$$
= 0.8022
$$

$$
\arctan = 38°44'09''
$$

$$
\text{bearing} = \text{N } 38°44'09'' \text{ W} \quad [\text{quadrant IV}]
$$

$$
\text{length} = \frac{\text{dep}}{\sin(\text{bearing})}
$$

$$
= 0.2119 \text{ ft}
$$

Check.

$$
\text{length} = \frac{\text{lat}}{\cos(\text{bearing})}
$$

$$
= 0.2119 \text{ ft}
$$

The total error of closure is N 38°44'09'' W, 0.2119 ft.

When the latitudes and departures have been corrected, new bearings and lengths for the course of the traverse need to be computed. Using the inversing Eqs. 6.3 through 6.5, compute the bearings and lengths of the courses using the corrected latitudes and departures.

The process of adjusting and balancing the traverse is complete when the bearings and distances of the courses are adjusted using the corrected latitudes and departures.

***Table 6.3** Corrected Traverse*

course	lat	dep	bearing $\tan(\text{bearing}) = \dfrac{\text{dep}}{\text{lat}}$	length $L = \dfrac{\text{dep}}{\sin(\text{bearing})} = \dfrac{\text{lat}}{\cos(\text{bearing})}$
AB	218.4651	313.9693	N 55°10'09'' E	382.497
BC	126.139	260.8157	S 64°11'24'' E	289.717
CD	455.701	280.5885	S 31°37'19'' W	535.157
DA	363.3748	294.1965	N 38°59'40'' W	467.539

C. Balancing an Open Traverse

An open traverse can be balanced if the coordinate values of the end or closing point are known. The procedure is similar to the compass method adjustment. Instead of adjusting the latitudes and departures, the coordinates of the traverse points are adjusted directly. The coordinates of the points of the traverse are computed by the method shown in Ex. 6.4. The difference between the coordinates derived from the traverse and the known coordinates is computed. The correction to the unadjusted northing is computed using Eq. 6.6.

$$C_{\rm N} = \left(\frac{d_{\rm N}}{L}\right)(l) \qquad 6.6$$

$C_{\rm N}$ = correction to northings
$d_{\rm N}$ = difference between ending and known northings
L = overall length of the traverse
l = distance from traverse point to beginning point

The corrections to the unadjusted eastings is computed using Eq. 6.7.

$$C_{\rm E} = \left(\frac{d_{\rm E}}{L}\right)(l) \qquad 6.7$$

$C_{\rm E}$ = correction to eastings
$d_{\rm E}$ = difference between ending and known eastings
L = overall length of the traverse
l = distance from traverse point to beginning point

Example 6.5

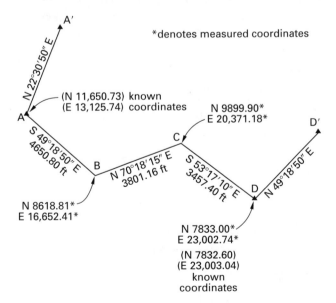

*denotes measured coordinates

A'
N 22°30'50" E
(N 11,650.73) known
(E 13,125.74) coordinates
A
S 49°18'50" E
4650.80 ft
B
N 8618.81*
E 16,652.41*
N 70°18'15" E
3801.16 ft
C
N 9899.90*
E 20,371.18*
S 53°17'10" E
3457.40 ft
D
N 7833.00*
E 23,002.74*
(N 7832.60)
(E 23,003.04)
known
coordinates
N 49°18'50" E
D'

The above illustration shows an open traverse. The angles and distances have been measured between points A and D, through points B and C. The coordinates of points A and D are known. The angles have

been adjusted to fit the bearings of lines A-A' and D-D'. The coordinate values (*) of points B, C, and D have been computed using the methods outlined in this chapter. Compute the adjusted coordinate values of points B and C.

Solution

Compute the difference between the ending coordinate and the known coordinate value of point D.

$$d_{\rm N} = 7833.00 \text{ ft} - 7832.60 \text{ ft} = 0.40 \text{ ft}$$
$$d_{\rm E} = 23{,}003.04 \text{ ft} - 23{,}002.74 \text{ ft} = 0.30 \text{ ft}$$

Compute the total and relative error of closure.

$$\tan(\text{bearing}) = \frac{\text{dep}}{\text{lat}} = \frac{0.30 \text{ ft}}{0.40 \text{ ft}}$$
$$\arctan = 36°52'12'' = \text{S } 36°52'12'' \text{ E}$$
$$l = \frac{\text{dep}}{\sin(\text{bearing})} = 0.50'$$

Check.

$$l = \frac{\text{lat}}{\cos(\text{bearing})} = 0.50'$$

The relative error of closure is the total error in relation to the overall length.

$$\frac{0.50'}{11{,}909.36} = \frac{1}{24{,}000} \quad [\text{approximately } 1{:}24{,}000]$$

Compute the corrected northings of points B and C. The northings of the ending point are too large; therefore, the corrections are subtracted from the unadjusted coordinates.

The corrections to point B northings are

$$C_{\rm N} = \left(\frac{d_{\rm N}}{L}\right)(l)$$
$$= \left(\frac{0.40 \text{ ft}}{11{,}909.36 \text{ ft}}\right)(4650.80 \text{ ft})$$
$$= 0.16 \text{ ft}$$
$$\text{northing B} = 8618.81 \text{ ft} - 0.16 \text{ ft} = 8618.65 \text{ ft}$$

The corrections to point C northings are

$$C_{\rm N} = \left(\frac{d_{\rm N}}{L}\right)(l)$$
$$= \left(\frac{0.40 \text{ ft}}{11{,}909.36 \text{ ft}}\right)\left(\begin{array}{l} 4650.80 \text{ ft} \\ + 3801.16 \text{ ft} \end{array}\right)$$
$$= 0.28 \text{ ft}$$
$$\text{northing C} = 3899.90 \text{ ft} - 0.28 \text{ ft} = 9899.62 \text{ ft}$$

Compute the corrected eastings of points B and C. The eastings of the ending point are too small; therefore, the corrections are added to the unadjusted coordinates.

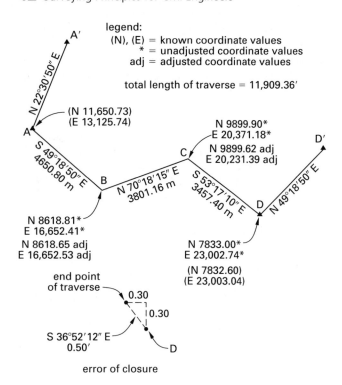

legend:
(N), (E) = known coordinate values
* = unadjusted coordinate values
adj = adjusted coordinate values

total length of traverse = 11,909.36'

(N 11,650.73)
(E 13,125.74)

N 9899.90*
E 20,371.18*
N 9899.62 adj
E 20,231.39 adj

N 22°30'50" E

A

S 49°18'50" E
4650.80 m

B

N 70°18'15" E
3801.16 m

C

S 53°17'10" E
3457.40 m

D

N 49°18'50" E

N 8618.81*
E 16,652.41*
N 8618.65 adj
E 16,652.53 adj

N 7833.00*
E 23,002.74*

(N 7832.60)
(E 23,003.04)

end point
of traverse

0.30

0.30

S 36°52'12" E
0.50'

D

error of closure

The corrections to point B eastings are

$$C_E = \left(\frac{d_E}{L}\right)(l)$$

$$= \left(\frac{0.30 \text{ ft}}{11,909.36 \text{ ft}}\right)(4650.80 \text{ ft})$$

$$= 0.12 \text{ ft}$$

easting B = 16,652.41 ft + 0.12 ft = 16,652.53 ft

The corrections to point C eastings are

$$C_E = \left(\frac{d_E}{L}\right)(l)$$

$$= \left(\frac{0.30 \text{ ft}}{11,909.36 \text{ ft}}\right)(4650.80 \text{ ft} + 3801.16 \text{ ft})$$

$$= 0.21 \text{ ft}$$

easting C = 20,231.18 ft + 0.21 ft = 20,231.39 ft

After the coordinates are adjusted, inverses are performed between the consecutive coordinates to adjust the bearings and lengths of the courses accordingly. The following table shows the final adjustments to the coordinates and the courses.

point	bearing	length	northing	easting
A			11,650.73	13,125.74
	S 49°18'48" E	4650.99		
B			8618.65	16,652.53
	N 70°18'23" E	3801.20		
C			9899.62	20,231.39
	S 53°17'08" E	3457.54		
D			7832.60	23,003.04

5. Rectangular Coordinates

The *rectangular coordinates* of a point on a traverse can be defined as "the perpendicular distances (coordinates) of a point from a pair of axes which intersect at right angles, reckoned in the plane defined by those axes" (*ACSM—Definitions of Surveying and Associated Terms*). The rectangular components of a line are the latitude and departure of that line. To compute the rectangular coordinates of any point on a line or traverse, add or subtract, depending on the direction, the latitudes and departures to or from the coordinates of the origin point. North latitudes and east departures are added, and south latitudes and west departures are subtracted.

Example 6.6

Using the traverse shown in Ex. 6.4 and the values from Table 6.3, compute the coordinates for the points of the traverse. The starting coordinates, at point A, are N 10,000.00, E 10,000.00.

Solution

This example is best solved in tabular form.

From the following table, the coordinate value of a point is shown as

$$y \text{ (northing)}, \quad x \text{ (easting)}$$

point	lat +(N)	lat −(S)	dep +(E)	dep −(W)	coordinates y	coordinates x
A					10,000.00	10,000.00
	218.47		313.97		+218.47	+313.97
B					10,218.47	10,313.97
		126.14	260.82		−126.14	+260.82
C					10,092.33	10,574.79
		455.70		280.59	−455.70	−280.59
D					9636.63	10,294.20
	363.37			294.20	+363.37	−294.20
A					10,000.00	10,000.00

The *y*-values are derived from the latitudes and the *x*-values from the departures. The last calculation shown on the table is the computation from the next to the last point (D) back to the starting point (A). In a balanced traverse, the starting and ending coordinates will be the same.

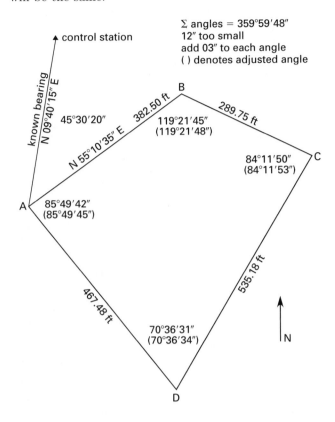

Figure 6.5 *Errors of Closure*

6. Enclosed Areas

When a parcel of land has been traversed, its area can be computed using the latitudes and departures of the courses in the traverse. The traverse must be balanced prior to computing the area. The method most commonly used to compute areas of a closed traverse is called the *double meridian distance* (DMD) method. The DMD method uses a series of triangles and trapezoids that are both included and excluded from the traverse.

A. Double Meridian Distances

The meridian distance of a course is the perpendicular distance to the midpoint of a course from the reference line or *y*-axis. The DMD is twice the meridian distance. DMDs are used to simplify the computation as this method uses the formulas for the area of trapezoids and triangles employing one-half the base. Instead of dividing by two each time a calculation is made, the DMD is used and the result is then divided by two.

The meridian distances for each course of traverse ABCDA are shown in Fig. 6.6. When area computations are made, the algebraic signs of the departures must be used. To avoid using negative values, the reference meridian is drawn through the most westerly point of the traverse.

The following rules for computing DMDs can be made.

- The DMD of the first course is equal to the departure of itself.

- The DMD of all following courses equals the DMD of the previous course plus the departure of the previous course plus the departure of itself.

- The DMD of the final course will be the same as the departure of itself with an opposite sign.

- The sign of all DMDs is positive.

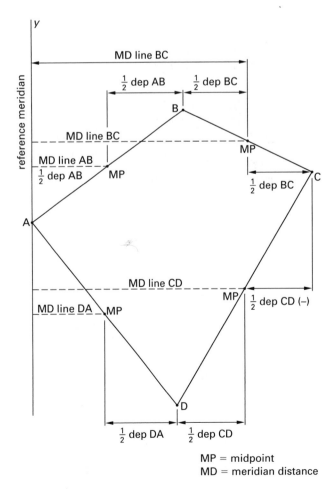

Figure 6.6 *Meridian Distances*

B. Area Computation Using DMD

The double area of a traverse is computed in the following manner.

step 1: Multiply the DMD of each course by the latitude of the course. North latitudes are positive and south latitudes are negative.

step 2: Compute the algebraic sum of the double areas.

step 3: Divide the algebraic sum of the double areas by two, thus yielding the area of the traverse in square feet.

step 4: Divide the results by 43,560 to obtain the area in acres.

Example 6.7

Using the data from Table 6.5, compute the area included in traverse ABCDA. Round the values of the latitudes and departures to the nearest 0.01′.

Solution

(A) Compute the DMDs of each course.

dep for AB $=$ $+313.97$ $=$ DMD for AB

$$\begin{array}{rl} \text{dep for AB} & = +313.97 \\ \text{dep for BC} & = \underline{+260.82} \\ & 888.76 = \text{DMD for BC} \end{array}$$

$$\begin{array}{rl} \text{dep for BC} & = +260.82 \\ \text{dep for CD} & = \underline{-280.59} \\ & 868.99 = \text{DMD for CD} \end{array}$$

$$\begin{array}{rl} \text{dep for CD} & = -280.59 \\ \text{dep for DA} & = \underline{-294.20} \\ & 294.20 = \text{DMD for DA} \end{array}$$

The DMD for the final course, DA, is equal to the departure of DA but with an opposite sign.

(B) The following table shows the double area of each course.

7. Adjustments to Latitudes and Departures

The three most common methods of adjusting a traverse such that it balances are the transit rule, the compass rule, and least squares.

The *transit rule* is used when the angular measurements are judged to be more reliable (or accurate) than the linear (distance) measurements. The error is distributed in proportion to the magnitude of the latitude and departure of each course. In Ex. 6.4, the error of closure of latitudes was computed to be 0.1653 ft, which would be distributed among courses in proportion to the latitude of each course. For course AB, the correction would be 218.4274 ft/(581.7561 ft + 581.9214 ft) = 0.1877; that is, 18.77% of the 0.1653 ft error would be assigned to course AB. In other words, the latitude of course AB would be adjusted by 18.77% of 0.1653 ft.

The *compass rule* is used when the angular and linear measurements are judged to be of equal reliability. The error is distributed to each course in proportion to the leg distances. In Ex. 6.4, the error of closure is again 0.1653 ft in latitude, so for course AB, the correction to its latitude would be 382.50 ft/(382.50 ft + 289.75 ft + 535.18 ft + 467.48 ft) = 0.2283, or 22.83% of the 0.1653 error. This method is regarded as superior to the transit rule.

The method of *least squares* can be applied to any traverse regardless of the relative reliability of the angular and linear measurements. It involves first calculating the slope of a straight line drawn through the graphed data points (field measurements), and then determining the goodness of fit of this line. Such curve-fitting techniques are usually incorporated into various surveying computer software packages available today, and they are difficult to perform by hand.

The compass rule will be used to adjust the traverse in Ex. 6.4.

course	lat (sign)	dep (sign)	DMD	double area (DMD× lat)		
			(+)	(+)	(−)	
AB	(+)218.47	(+)313.97	313.97	68,593.03		
BC	(−)126.14	(+)260.82	888.76		112,108.19	
CD	(−)455.70	(−)280.59	868.99		395,998.74	
DA	(+)363.37	(−)294.20	294.20	106,903.45		
				175,496.48	508,106.93	
					175,496.48	
			double area	= 332,610.45		
			divide by 2	= 166,305.23 ft^2		

$$\text{area} = \frac{166{,}305.25 \text{ ft}^2}{43{,}560 \, \dfrac{\text{ft}^3}{\text{ac}}} = 3.82 \text{ ac}$$

To correct the latitude of a course,

$$C_{\text{lat}} = d_{\text{lat}} \left(\frac{l}{p} \right) \qquad 6.8$$

C_{lat} = correction to latitude of the course

d_{lat} = difference between summed latitudes

p = perimeter of traverse
(equal to the sum of the headings)

l = length of an individual course
(or an individual course heading)

To correct the departure of a course,

$$C_{\text{dep}} = d_{\text{dep}} \left(\frac{l}{p} \right) \qquad 6.9$$

C_{dep} = correction to departure of the course

d_{dep} = difference between summed departures

p = perimeter of traverse

l = length of an individual course

The corrections are computed and then applied to the latitudes and departures of each course separately in the following manner.

- If the northings exceed the southings, subtract the correction assigned to it from the north latitudes and add to the south latitudes.

- If the southings exceed the northings, add the correction assigned to it to the north latitudes and subtract from the south latitudes.

- If the eastings exceed the westings, subtract the correction assigned to it from the east departures and add to the west departures.

- If the westings exceed the eastings, add the correction assigned to it to the east departures and subtract from the west departures.

Northings are positive latitudes, southings are negative latitudes, eastings are positive departures, and westings are negative departures. In order for the traverse in Fig. 6.6 to close on point A, the latitudes need to be increased and the departures need to be decreased.

Returning to Ex. 6.4, compute and apply corrections to the latitudes and departures of the traverse such that it balances.

Table 6.5 shows the results of the computation and application of the corrections to the latitudes and departures. The algebraic sum of the latitudes and the algebraic sum of the departures are both equal to zero; therefore, the traverse now closes.

Table 6.4 *Corrections to Latitudes and Departures*

course	length	correction to lat	correction to dep
AB	382.50	$\left(\frac{0.1653}{1675}\right)(382.50) = 0.0377$	$\left(\frac{0.1326}{1675}\right)(382.50) = 0.0303$
BC	289.75	$\left(\frac{0.1653}{1675}\right)(289.75) = 0.0286$	$\left(\frac{0.1326}{1675}\right)(289.75) = 0.0229$
CD	535.18	$\left(\frac{0.1653}{1675}\right)(535.18) = 0.0528$	$\left(\frac{0.1326}{1675}\right)(535.18) = 0.0424$
DA	467.48	$\left(\frac{0.1653}{1675}\right)(467.48) = 0.0461$	$\left(\frac{0.1326}{1675}\right)(467.48) = 0.0370$

Table 6.5 *Corrected Latitudes and Departures*

course	lat	correction $\left(\frac{0.1653}{1675}\right)(l)$	adjusted lat	dep	correction $\left(\frac{0.1326}{1675}\right)(l)$	adjusted dep
AB	218.4274	0.0377	218.4651	313.9996	−0.0303	313.9693
BC	−126.1676	−0.0286	−126.139	260.8386	−0.0229	260.8157
CD	−455.7538	−0.0528	−455.701	−280.5461	0.0424	−280.5885
DA	363.3287	0.0461	363.3748	−294.1595	0.0370	−294.1965
			sum = 0			sum = 0

Practice Problems

Problems 1–4 refer to the following illustration which represents the results of a field survey and has not been adjusted.

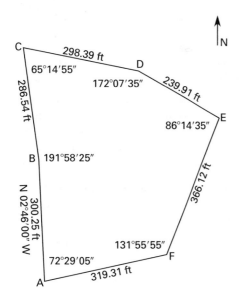

1. Balance the traverse and compute the adjusted bearings and distances for each course.

2. Using the adjusted data from Prob. 1, compute the area of figure ABCDEFA using the DMD method.

3. Using the adjusted data from Prob. 1, compute the coordinates of points B, C, D, E, and F. The coordinates of point A are N 10,000; E 5000.00.

4. Compute the bearing and distance from point B to point E.

Solutions

1. Follow the procedures outlined in steps 1 through 7 in Sec. 4.B.

step 1: Adjust field angles.

Compute the sum of the interior angles.

A	72°29′05″
B	191°58′25″
C	65°14′55″
D	172°07′35″
E	86°14′35″
F	131°55′55″
sum	720°00′30″

The sum should equal $(n-2)(180°)$. n is the number of angles; therefore, the sum should equal $(6-2)(180°) = 720°$. The sum of the angles is 30″ too large. Since there are six angles, make each one 05″ smaller.

The adjusted angles are

A	72°29′00″
B	191°58′20″
C	65°14′50″
D	172°07′30″
E	86°14′30″
F	131°55′50″
sum	720°00′00″ $= (n-2)(180°)$
	$= (4)(180°)$
	$= 720°$

step 2: Compute the bearings of each course.

Using the methods described in Chap. 4, compute the bearings of each course.

course AB: N 02°46′00″ W is given.

course BC: 191°58′20″ − 180° = 11°58′20″
02°46′00″ + 11°58′20″ = 14°44′20″
N 14°44′20″ W

course CD: 14°44′20″ + 65°14′50″ = 79°59′10″
S 79°59′10″ E

course DE: 180° − 172°07′30″ = 07°52′30″
79°59′10″ − 07°52′30″ = 72°06′40″
S 72°06′40″ E

course EF: 72°06′40″ + 86°14′30″ = 158°21′10″
180° − 158°21′10″ = 21°38′50″
S 21°38′50″ W

course FA: 180° − 131°55′50″ = 48°04′10″
48°04′10″ + 21°38′50″ = 69°43′00″
S 69°43′00″ W

Using angle A, compute the bearing of course AB as a check.

course AB: 72°29′00″ − 69°43′00″ = 02°46′00″
N 02°46′00″ W [OK]

steps 3 and 4: Compute the latitudes, departures, and misclosures of the courses.

$$\text{lat} = \cos(\text{bearing}) \times (\text{length})$$
$$\text{dep} = \sin(\text{bearing}) \times (\text{length})$$

course	bearing (D°M′S″)	length (ft)	north lat (ft)	easting dep (ft)
AB	N 02°46′00″ W	300.25	299.90	−14.49
BC	N 14°44′20″ W	286.54	277.11	−72.90
CD	S 79°59′10″ E	298.39	−51.88	293.84
DE	S 72°06′40″ E	239.91	−73.70	228.31
EF	S 21°38′50″ W	366.12	−340.30	−135.06
FA	S 69°43′00″ W	319.31	−110.69	−299.50
sum		1810.52	0.44	0.20

The traverse miscloses by +0.44 ft in latitudes and +0.20 ft in departures. Consequently, to achieve closure, (+) latitudes must be reduced, (−) latitudes increased, (+) departures reduced, and (−) departures increased. The size of corrections are proportioned by applying the compass rule adjustment.

step 5: Compute the error of closure.

The tangent of the bearing of the error of closure is

$$\frac{\Delta \text{ dep}}{\Delta \text{ lat}} = \frac{0.20 \text{ ft}}{0.44 \text{ ft}} = 0.454545$$

$$\arctan = 24°26'38'' = \text{S } 24°26'38'' \text{ W}$$

$$\text{length} = \frac{\text{dep}}{\sin(\text{bearing})} = 0.48 \text{ ft}$$

Check.

$$\text{length} = \frac{\text{lat}}{\cos(\text{bearing})} = 0.48 \text{ ft} \quad \text{[OK]}$$

The total error of closure is

$$\text{S } 24°26'38'' \text{ W, } 0.48 \text{ ft}$$

The relative error of closure is

$$\frac{\text{total error of closure}}{\text{length of traverse}} = \frac{0.48 \text{ ft}}{1810 \text{ ft}}$$

Therefore, the relative error of closure is 1:3770.

step 6: Compute the corrections to the latitudes and departures.

Use the compass rule adjustment. Compute the correction factor for the latitudes.

$$C_{\text{lat}} = \frac{d_{\text{lat}}}{p} = \frac{0.44 \text{ ft}}{1810.52 \text{ ft}} = 0.000243 \text{ ft}$$

Compute the correction factor for the departures.

$$C_{\text{dep}} = \frac{d_{\text{dep}}}{p} = \frac{0.20 \text{ ft}}{1810.52 \text{ ft}} = 0.000110 \text{ ft}$$

The corrections to the latitudes and departures of each course are computed by multiplying the length of the course by the correction factor computed previously.

The sum of the positive and negative latitudes and departures equal zero, thus the traverse is balanced.

step 7: Compute the bearings and distances of the courses from the adjusted latitudes and departures.

Compute the bearings.

$$\tan(\text{bearing}) = \frac{\text{dep}}{\text{lat}}$$

Compute the lengths.

$$\text{length} = \frac{\text{dep}}{\sin(\text{bearing})}$$

Check.

$$\text{length} = \frac{\text{lat}}{\cos(\text{bearing})}$$

course	adj bearing $\tan = \dfrac{\text{dep}}{\text{lat}}$	adj length $\dfrac{\text{dep}}{\sin(\text{brg})}$; $\dfrac{\text{lat}}{\cos(\text{brg})}$	adj lat (ft)	adj dep (ft)
AB	N 02°46'21'' W	300.18	299.83	−14.52
BC	N 14°44'54'' W	286.48	277.04	−72.93
CD	S 79°58'22'' E	298.37	−51.95	293.81
DE	S 72°05'37'' E	239.90	−73.76	228.28
EF	S 21°38'53'' W	366.22	−340.29	−135.10
FA	S 69°42'20'' W	319.37	−110.77	−299.54

course	length (*l*)	lat corr (*l*)(0.000243)	dep corr (*l*)(0.000110)	adj lat (ft)	adj dep (ft)
AB	300.25	−0.07	0.03	299.83	−14.52
BC	286.54	−0.07	0.03	277.04	−72.93
CD	298.39	0.07	−0.03	−51.95	293.81
DE	239.91	0.06	−0.03	−73.76	228.28
EF	366.12	0.09	0.04	−340.39	−135.10
FA	319.31	0.08	0.04	−110.77	−299.54
sum		0.44	0.20	0	0

2. From observation, point C is the most westerly point on the traverse; therefore, begin the DMD computations using course CD as the first course.

Calculate the DMDs.

$$\text{dep of CD} \quad +293.81 = \text{DMD for CD}$$

$$\begin{array}{ll} \text{dep of CD} & +293.81 \\ \text{dep of DE} & +228.28 \\ \hline & 815.90 = \text{DMD for DE} \end{array}$$

$$\begin{array}{ll} \text{dep of DE} & +228.28 \\ \text{dep of EF} & -135.10 \\ \hline & 909.08 = \text{DMD for EF} \end{array}$$

$$\begin{array}{ll} \text{dep of EF} & -135.10 \\ \text{dep of FA} & -299.54 \\ \hline & 474.44 = \text{DMD for FA} \end{array}$$

$$\begin{array}{ll} \text{dep of FA} & +299.54 \\ \text{dep of AB} & -14.52 \\ \hline & 160.38 = \text{DMD for AB} \end{array}$$

$$\begin{array}{ll} \text{dep of AB} & -14.52 \\ \text{dep of BC} & -72.93* \\ \hline & 72.93* = \text{DMD for BC} \end{array}$$

*The DMD of BC is equal to the departure of BC, but the signs are opposite.

Compute the area by DMD.

course	adj lat		DMD		double areas (+)	double areas (−)
CD	−51.95	×	293.81	=		15,263
DE	−73.76	×	815.90	=		60,181
EF	−340.39	×	909.08	=		309,442
FA	−110.77	×	474.44	=		52,554
AB	299.83	×	160.38	=	48,087	
BC	277.04	×	72.93	=	20,205	
			sum		68,292	−437,440
						+68,292
					double area	369,148 ft²

$$\text{area} = \frac{369{,}148 \text{ ft}^2}{2} = 184{,}574 \text{ ft}^2$$

$$\frac{184{,}574 \text{ ft}^2}{43{,}560 \ \dfrac{\text{ft}^3}{\text{ac}}} = 4.237 \text{ ac}$$

3.

point	lat (ft) (+)	lat (ft) (−)	dep (ft) (+)	dep (ft) (−)	coordinate y	coordinate x
A					10,000.00	5000.00
	299.83			14.52		
B					10,299.83	4985.48
	277.04			72.93		
C					10,576.87	4912.55
		51.95	293.81			
D					10,524.92	5206.36
		73.76	228.28			
E					10,451.16	5434.64
		340.39		135.10		
F					10,110.77	5299.54
		110.77		299.54		
A					10,000.00	5000.00

4. Subtract the northings.

$$10{,}451.16 \text{ ft} - 10{,}299.83 \text{ ft} = 151.33 \text{ ft} = \text{lat}$$

Subtract the eastings.

$$5434.64 \text{ ft} - 498.45 \text{ ft} = 449.16 \text{ ft} = \text{dep}$$

Compute the bearing.

$$\tan(\text{bearing}) = \frac{\text{dep}}{\text{lat}} = \frac{449.16 \text{ ft}}{151.33 \text{ ft}} = 2.96808$$

$$\arctan = 71°22'49''$$

The northings and eastings of point E are greater than those of point B; therefore the bearing is N 71°22'49″ E.

Solving for the length,

$$\text{length} = \frac{\text{dep}}{\sin(\text{bearing})} = 473.97 \text{ ft}$$

Check.

$$\text{length} = \frac{\text{lat}}{\cos(\text{bearing})} = 473.97 \text{ ft}$$

Topography 7

1. Introduction

Topography refers to the process of measuring and mapping the elevations and physical features of a site in plan view. Usually, these measurements and subsequent mapping are used for designing public or private infrastructure projects.

The field measurements involve any of the usual surveying instruments discussed in Chap. 1. Elevations, horizontal distances between a reference line or point to points of interest, and worded descriptions of any geologic, man-made, or natural features are taken for later plotting onto a map (also termed *mapping*).

The field measurements are organized using one of three basic methods: the cross-section method, the grid method, and the radial method. Mapping will look the same regardless of which method was used.

2. Basic Control

Prior to topographic surveying, both horizontal and vertical control must be established to allow features to be tied to some known points or lines. Once control is established, it can be used for the topographic survey, the design, and the construction phases of the project. The control datum must remain constant throughout these phases and must be reproduceable throughout the life of the project. *Horizontal control* can be property or boundary lines, street or highway centerlines, or random traverse lines tied to any of these or to a local or state plane coordinate system. *Vertical control* can be based on an assumed datum, a local datum, or the National Vertical Geodetic Datum (NVGD 1929) or the North American Vertical Datum of 1988 (NAVD 88). Most government agencies require that mapping be based on state plane coordinates and NVGD 1929. Many agencies are now specifying NAVD 88.

3. Cross-Section Method

In the *cross-section method*, cross sections for elevations are taken at right angles (offset lines) to stations along a baseline. Features such as trees, utilities, walls, fences, and so on are located at a right angle to the baseline from a station perpendicular to the feature. For public works projects such as street improvements or storm drain or channel work, the baseline is usually the existing or proposed centerline of the project. When an existing street or road is being improved, the centerline of that street or road makes the best baseline for cross-section survey. Often, a project is surveyed before the final or precise alignment is determined. In such a case, a random line is established in the vicinity of the proposed centerline and cross sections are taken from this line. The data gathered from this survey may be used to help determine the precise alignment. Once precise alignment is designed, a subsequent survey may be called for using the new alignment as a baseline; however, the second survey is usually not needed if the features located on the first survey can be related accurately to the new baseline.

4. Locating and Recording Features

In Fig. 7.1, the existing roadway centerline of Hughes Road is used as the horizontal control (baseline) for cross sectioning. The beginning station, 10+00, is at the intersection of Elm Street. Note that half-stations (at 50 ft intervals) are marked.

Each point of interest or feature is located both in the horizontal (the baseline) and in the vertical (its elevation). To fully describe a point in the horizontal, first determine the exact station where the point appears directly perpendicular from the baseline. This perpendicular line from the baseline is termed an *offset*. Second, measure the horizontal distance along the offset from the baseline to the point of interest. Thus, a point is described fully in the horizontal by noting both the station at which the offset to the point is and the offset distance to the point. The elevation of the point of interest completes the description.

Table 7.1 shows sample cross-section notes for the survey shown in Fig. 7.1. The stationing begins at the bottom of the page and increases toward the top of the page. Taking station 10+00 as an example, the first row of numbers (10.87, 10.53, 10.20, etc.) are rod readings taken to determine elevations at various points along the offset line perpendicular to station 10+00. The second row of numbers (50.0, 20.0, 0.0, etc.) indicate the offset distance from the baseline to the point. The point noted as 0.0 is the intersection of the offset line with the baseline. Thus, at a point on the offset line 20.0 ft to the left of the baseline, the rod reading is 10.53. At a point on the offset line 20.0 ft to the right of the baseline, the rod reading is 10.50. The rod readings will be reduced to elevations by subtracting them from the HI (height of instrument).

Several abbreviations are used in cross-section notes. EP means *edge of pavement* of a roadway, FL is *flow line* of a pipe, FG means *finished grade*, RW means *right of way* (a property line), and so on.

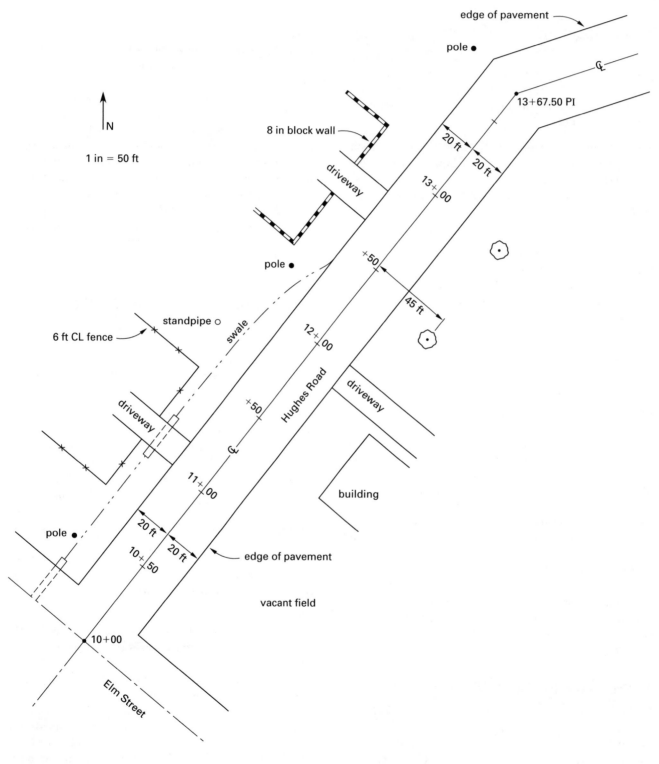

Figure 7.1 *Alignment Line Plan View*

Table 7.1 Cross-Section Notes

sta	description	left			Ȼ		right		
13+63	pole			X 36.0 pole	X 0.0 Ȼ				
13+50		7.3 50.0 FG	7.2 40.0 R/W	7.05 20.0 EP	6.70 0.0 Ȼ	7.00 20.0 EP	7.1 40.0 R/W	7.2 50.0 FG	
13+08	corner of wall			7.3 40.0 FG at corner wall	X 0.0 Ȼ				
13+05	22 in pine (diameter)		7.5 40.0 R/W	7.51 20.0 EP	7.10 0.0 Ȼ	7.43 20.0 EP	7.6 40.0 R/W	X 45.0 tree	7.6 50.0 FG
12+83	edge of concrete driveway	7.80 50.0	7.85 40.0	7.62 20.0 EP	7.22 0.0 Ȼ				
12+83	edge of concrete driveway	7.85 50.0	7.90 40.0	7.75 20.0 EP	7.40 0.0 Ȼ				
12+50			8.0 40.0 R/W	7.96 20.0 EP	7.56 0.0 Ȼ	7.90 20.0 EP	8.2 40.0	8.2 50.0	
12+28	corner of wall end swale	8.3 50.0 FG	2.40 40.0 TW	8.4 40.0 FG at corner wall	8.15 20.0 EP	7.80 0.0 Ȼ			
12+13	power pole #737E			X 35.0 pole	X 0.0 Ȼ				
12+4	18 in pine (diameter)			X 0.0 Ȼ	X 45.0 pine tree				
12+02	edge of concrete driveway	8.5 50.0 FG	8.6 40.0 R/W	9.4 26.0 FL	8.62 20.0 EP	8.10 0.0 Ȼ	8.60 20.0 EP	8.40 40.0 R/W	8.30 50.0
11+86	edge of concrete driveway				8.25 0.0 Ȼ	8.68 20.0 EP	8.48 40.0 R/W	8.36 50.0	
11+78	building corner			X 0.0 Ȼ	9.3 50.0 FG at bldg corner				
11+73	standpipe 18 in above ground			9.2 46.0 FG at std pipe	X 0.0 Ȼ				
HI = 1216.20 benchmark #37; el = 1205.80 BS = +10.40 HI = 1216.20									

(continued)

Table 7.1 (continued) *Cross-Section Notes*

sta	description	50 L	40 L	32 L	20 L	℄	20 R	40 R	50 R
11+47	corner—6 ft chain link fence		9.3 40.0 FG at fence corner			X 0.0 ℄			
11+36	building corner					X 0.0 ℄			9.4 50.0 FG at bldg corner
11+21	end 12 in concrete pipe		9.4 40.0 R/W	10.35 32.0 FL pipe	9.20 20.0 EP	8.82 0.0 ℄			
11+15	edge of asphalt driveway	9.18 50.0	9.05 40.0		9.25 20.0 EP	8.90 0.0 ℄			
								full section taken at 11+00	
10+66	corner—6 ft chain link fence	9.7 50.0 FG	9.9 40.0 FG at fence corner			X 0.0 ℄			
10+50		9.9 50.0 FG	10.1 40.0 R/W	10.9 32.0 FL	9.85 20.0 EP	9.50 0.0 ℄	9.90 20.0 EP	10.2 40.0 R/W	10.3 50.0 FG
10+40	power pole #735E		X 37.0 pole			X 0.0 ℄			
10+26.50	end 12 in concrete pipe		10.3 40.0 FG at RW	11.40 32.0 FL pipe	10.25 20.0 EP	9.88 0.0 ℄			
10+20	angle point edge of pavement	10.96 50.0			10.36 20.0 EP	10.00 0.0 ℄	10.40 20.0 EP		10.75 50.0
10+00	intersection Elm Street	10.87 50.0			10.53 20.0	10.20 0.0	10.50 20.0		10.95 50.0
				HI = 1216.20					
		benchmark #37; el = 1205.80 BS = +10.40 HI = 1216.20							

5. Grid Method

The *grid method* involves superimposing a grid onto a map of the area of interest.

First, the area to be mapped is divided into a series of equal squares by running lines on the ground parallel with and perpendicular to a convenient baseline. The elevations of all the corners of these squares and of intermediate high and low points are then determined. Next the points are plotted, and the contours are located on the plot from the observed elevations by interpolation. For convenience, each elevation is written on the plot so that the decimal point also marks the position of the point at which the elevation is determined.

To simplify the identification of the various points in the notes, it is customary to designate by letters the division lines that extend in one direction, and by numerals the lines that run in the perpendicular direction. The point at the intersection of any two lines is then

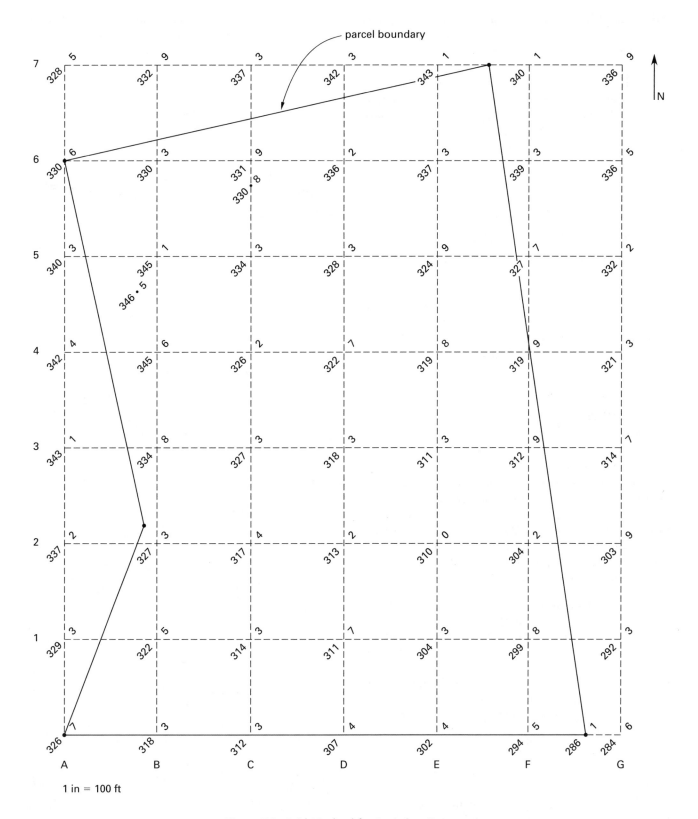

Figure 7.2 *Grid Method for Locating Contours*

designated by the letter and numeral of the respective intersecting lines. In Fig. 7.2, the line AG is selected as a base. The division lines that are perpendicular to the base are designated as lines A, B, C, and so on; and the lines parallel with the base are designated as lines 1, 2, 3, and so on. The intersection of line D and line 5 is designated D5. If the low point between C5 and C6, whose elevation is 330.8 ft, is 75 ft from C5, it is designated as C5+75. Also, the high point at elevation 346.5 ft, which is located between lines 4 and 5 and A and B, is called A+80, 4+60. In this method the intersection points are located at definite horizontal intervals and, consequently, those points do not generally lie on the contours. The points on the contours are then plotted by assuming that the ground slopes uniformly between the points whose elevations have been established.

6. Radial Topography Survey

A *radial topography survey* is made when the positions of the features or points are located by measuring angles and distances from a known or coordinated station and baseline. The two commonly used methods are the *transit-stadia* method and the *total station* method. Both methods allow for direction, distance, and elevation to be measured simultaneously.

7. Contours

A. Introduction

After field measurements have been taken and the survey notes have been reduced (or corrected), the data can be mapped onto a topographic map. Such mapping can be in the form of either spot elevations or contour lines. *Spot elevations* involve simply plotting each data point, with descriptions, onto a plan view map. *Contours*, however, involve plotting lines of equal elevations and subsequently drawing a smooth line through the points.

Properly located contours indicate elevations with a relatively high degree of accuracy. Also, contours can be drawn on a topographic map easily and rapidly and are more commonly used than other symbols for showing map relief. The United States Coast and Geodetic Survey, the United States Geological Survey, and many engineers engaged in private enterprises show practically all map relief by means of contours. A map whose primary purpose is to indicate the positions of contours is a *contour map* (see Fig. 7.3).

B. Elevations

Elevations are shown on a contour map by marking the numerical values of the elevations in feet on some of the contour lines and also at high and low points.

To make the configuration of the ground readily visible from a contour map, the vertical interval between adjacent contours (the difference between the elevations) is made the same for all parts of the map. This

vertical interval on a particular map is the *contour interval* for that map. The most legible contour interval depends on the degree of accuracy desired, the slope of the ground, and the scale of the map.

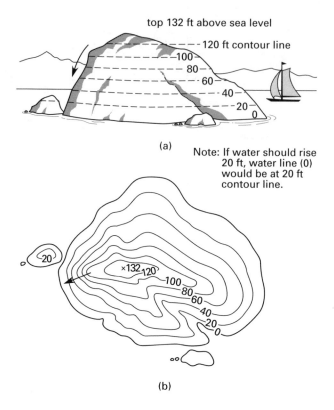

Note: If water should rise 20 ft, water line (0) would be at 20 ft contour line.

Figure 7.3 Contour Map

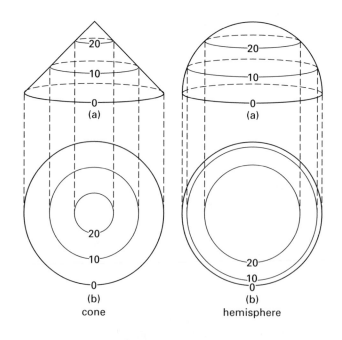

Figure 7.4 Contours

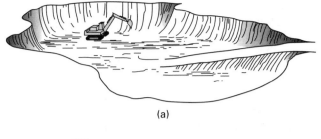

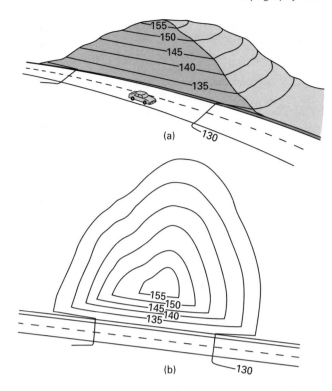

(a)

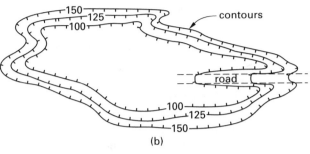

Figure 7.5 *Depression Contours*

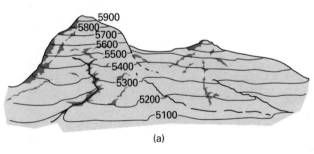

Figure 7.7 *Cut Section*

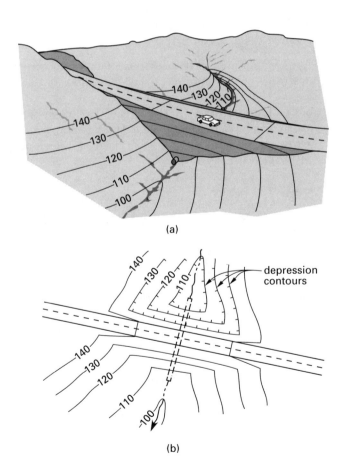

Figure 7.6 *Fill Section*

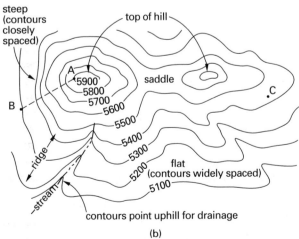

Figure 7.8 *Saddle Contours*

C. Representation of Typical Formations by Means of Contours

Figures 7.4 through 7.8 represent the five most common formations on the earth's surface. In each illustration, the surface formation is represented in view (a) by means of a perspective and in view (b) by means of contours on a plan. Arrows indicate the direction in which water would flow (that is, the direction in which the ground slopes downward), but such arrows would not be shown on an actual contour map.

For a hill or an island, each contour has the general form of a loop as shown in Fig. 7.4. The higher contours are smaller in perimeter than the lower contours and are entirely enclosed by the lower ones. The contour lines are drawn freehand and are made somewhat smooth to conform more nearly to the irregularities of the ground surface. The ground represented in this case is regular, but the contours for more rugged territory would differ only in the size of the indentations in the loops.

Contours at a depression or hollow in the ground have the same general appearance as contours at a hill. Such formations of fairly irregular shape are represented in Fig. 7.5. In the case of a depression, the lower contours are smaller than, and enclosed by, the higher ones.

If both a hill and a depression are shown on a contour map, it is customary to make an obvious distinction between the hill and the depression by drawing short lines perpendicular to the contours at the depression as in Fig. 7.5(b).

Figure 7.6 represents a highway fill. Here, a depression has been filled with dirt from a highway cut. Fig. 7.6(b) illustrates a contour map of a fill section.

Figure 7.7 represents a highway cut. Here, a portion of a hill has been excavated to accommodate a highway. Where the ground is steep, like the cut near the road, the contours are close together. Where the ground is flatter, the contours are farther apart.

Figure 7.8 illustrates various topographic features, such as a mountain saddle, a ridge, a creek, and flat areas.

D. Characteristics of Contours

Contours have the following significant characteristics.

- All points on a given contour have the same elevation.

- A contour on the ground closes on itself. A contour may close within a map as indicated in Figs. 7.4 and 7.5, or it may be discontinued at any two points at the borders of the map as indicated in Figs. 7.6 and 7.7. Such points mark the limits of the contour on the map, but the contour does not end at those points.

- The direction of a contour at any point is at right angles to the direction of the steepest slope of the ground at that point.

- Contours on the ground cannot cross one another, nor can contours having different elevations come together and continue as one line. However, where an overhanging cliff or a cave is represented on a map, contours on the map may cross. The lower contour must then be shown dotted. At a vertical ledge or wall, two or more contours may merge.

- The horizontal distance between adjacent contours indicates the steepness of the slope of the ground between the contours. Where the contours are relatively close together, the slope is comparatively steep, and where the contours are far apart the slope is gentle. For example, the surface of the hemisphere in Fig. 7.4 slopes more steeply between elevations 0 and 10 than between elevations 10 and 20, and in view (b) the horizontal distance between contours 0 and 10 is less than the distance between contours 10 and 20. Also, where the spaces between contours are equal, the slope is uniform, and where the spaces are unequal, the slope is not uniform. Thus, in the case of the cone in Fig. 7.4, the curved surface slopes uniformly from the base to the apex, and the contours in view (b) are spaced equally. In Figs. 7.6 and 7.7, variable slopes are illustrated.

- As a contour approaches a stream, the contour turns upstream until it intersects the shore line. It then crosses the stream and turns back along the opposite bank of the stream. If the stream has an appreciable width on the map, the contour is not drawn across the stream but is discontinued at the shore with which it merges.

E. Slope

The *slope* or gradient of the land can be determined from a contour map by dividing the difference in elevation (as determined from the contours) by the length of the line between the known points. Suppose the average gradient between points A and B on Fig. 7.8 is to be calculated. Assuming that the map is drawn at a scale of 1 in = 600 ft, the elevation difference is found by subtracting the elevation at A from the elevation at B.

$$\text{elev A} - \text{elev B} = 5900 \text{ ft} - 5300 \text{ ft} = 600 \text{ ft}$$

Scaling between points A and B yields a distance of 930 ft. The gradient between A and B is

$$S = \frac{600 \text{ ft}}{930 \text{ ft}}$$
$$= 0.645 \quad (64.5\%)$$

F. Profile

The *profile* is determined by scaling the distance between contour lines and plotting the horizontal distance and difference in elevations on a profile diagram. For line AB in Fig. 7.8, the vertical difference is 600 ft and the horizontal distance is 930 ft. Figure 7.9 shows the profile of the line AB using a scale of 1 in = 100 ft vertical + horizontal.

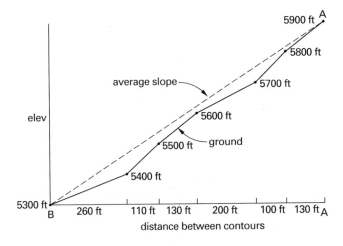

Figure 7.9 *Plotting a Profile from Contours*

G. Interpolation

Interpolation is used to compute the elevation of a feature or spot between contour lines or to compute the location of contour lines between spot elevations.

Example 7.1

At the feature located at point C in Fig. 7.8, there is no elevation specifically shown. Compute the elevation by interpolation.

Solution

Construct a line, normal to the 5400 ft contour, through point C and extended to the 5500 ft contour.

Scale the distance between the 5400 ft contour line, point C, and the 5500 ft contour line using a scale of 1 in = 600 ft.

$$5400 \text{ ft to C} = 80 \text{ ft}$$
$$\text{C to } 5500 \text{ ft} = 250 \text{ ft}$$
$$\text{overall distance} = 330 \text{ ft}$$

$$\frac{x}{80 \text{ ft}} = \frac{100 \text{ in}}{330 \text{ ft}}$$

x is the difference in elevation from the 5400 ft line to point C.

$$(330 \text{ ft})x = 8000 \text{ ft}$$
$$x = 24 \text{ ft}$$
$$\text{elevation C} = 5400 \text{ ft} + 24 \text{ ft}$$
$$= 5424 \text{ ft}$$

Practice Problems

1. Use the data given in the following cross-section notes to prepare a 10 ft contour interval plot at 1 in = 50 ft.

Cross-Section Notes

	left			₵	right		
sta	150	100	50	0	50	100	150
14+00	237.5	242.5	247.5	252.5	257.5	262.5	265.0
13+00	242.5	247.5	252.5	257.5	262.5	265.0	275.0
12+00	247.5	252.5	257.5	262.5	265.0	275.0	285.0
11+00	252.5	257.5	262.5	265.0	275.0	285.0	295.0
10+00	257.5	262.5	265.0	275.0	285.0	295.0	296.0

2. Use the data given in the following cross-section notes to prepare a 4 ft contour interval plot at 1 in = 50 ft. All sections are taken only to the left of the baseline.

Cross-Section Notes

sta	top	toe	top ditch	FL ditch	top ditch	baseline
18+00	258.6	233.4	232.7	228.5	232.7	236.0
	318.0	185.0	165.0	140.0	107.0	0.0
17+00	256.0	230.0	228.0	224.6	228.0	231.0
	315.0	170.0	140.0	115.0	85.0	0.0
16+00	253.0	227.0	225.0	221.0	225.0	227.2
	312.0	153.0	118.0	92.0	63.0	0.0
15+00	250.0	222.6	220.0	216.5	220.0	222.0
	309.0	137.0	95.0	70.0	40.0	0.0

3. Which of the following statements relating to contours is true?

    (A) All points on a contour line have the same elevation.

    (B) Contour lines close on themselves.

    (C) The direction of a contour line at any given point is at right angles to the direction of the steepest slope at that point.

    (D) All of the above are true.

4. What is a single line, such as a centerline or random line, that shows stations and elevation along its length?

(A) cross section

(B) contour line

(C) profile line

(D) control line

5. A street is to be widened and a topo survey is needed for design purposes. Normally, what survey method would be used?

(A) cross section

(B) grid

(C) photogrammetry

(D) radial topography

6. A straight line intersects two 20 ft contours, and the distance between the intersection points is 320 ft. What is the slope of line?

(A) 0.0625

(B) 6.25/100

(C) 16:1

(D) all of the above

7. What does the following contour pattern indicate?

(A) ridge

(B) hill

(C) stream or creek

(D) saddle

8. What does the following contour pattern indicate?

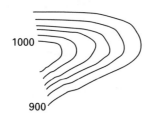

(A) ridge

(B) hill

(C) stream or creek

(D) saddle

9. An EDM is set on sta A whose elevation is 150.25 ft. The HI is measured as 5.20 ft. The zenith angle to the prism at sta B is $88°15'$, and the slope distance is measured as 765.85 ft. The height of the prism (HT) is 7.60 ft. What is the elevation of sta B?

(A) 124.47 ft

(B) 160.83 ft

(C) 171.24 ft

(D) 186.43 ft

Solutions

1. *step 1:* Plot the cross-section data on a grid at a scale of 1 in = 50 ft.

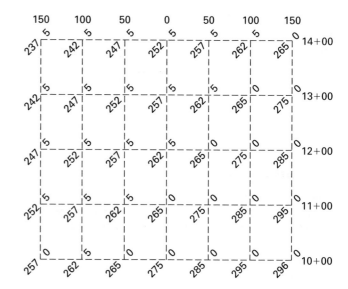

step 2: Plot the location of the even 10 ft contours between the elevations plotted on the grid by interpolation and connect the contour points with a solid line.

Locate the 260 ft contour line along sta 11+00. By observation, this contour is located between the 50 ft and 100 ft spot elevations. Set up a ratio to determine the slope between those shots. The difference in elevation is 262.5 ft − 258.3 ft = 4.0 ft. The ratio is

$$\frac{4 \text{ ft}}{50 \text{ ft}} = \frac{1.5 \text{ ft}}{x}$$

x is the difference from the 258.5 ft spot elevation, and 1.5 ft is the difference in elevation.

$$260 \text{ ft} - 258.5 \text{ ft} = 1.5 \text{ ft}$$

Solve for x.

$$4x = 75 \text{ ft}$$
$$x = 18.75 \text{ ft}$$

Plot the 260 ft contour 18.75 ft from the 258.5 ft spot, toward the 262.5 ft spot.

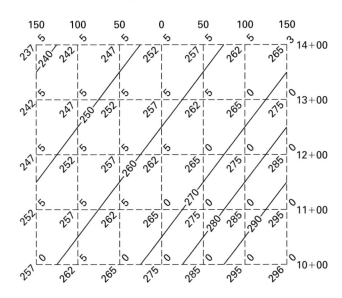

step 3: Locate the even contours between the slope elevations and draw lines between them.

2. See the following illustration.

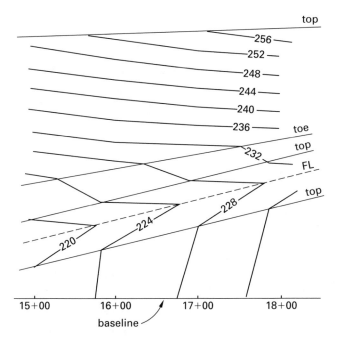

3. Answer (D)

4. Answer (C)

5. Answer (A)

6.
$$\text{slope} = \frac{\text{difference in elevation}}{\text{horizontal distance}} = \frac{20 \text{ ft}}{320 \text{ ft}} = \frac{1}{16}$$
$$= 0.0625$$

Answer (D)

7. Answer (C)

8. Answer (A)

9. Solve for V.
$$V = D(\cos Z)$$
$$= (765.85 \text{ ft})(\cos 88°15')$$
$$= 23.39 \text{ ft}$$

$$\text{elevation at station B} = \text{elev sta A} + \text{HI} + V(Z > 90°) - \text{HT}$$
$$= 150.25 \text{ ft} + 5.20 \text{ ft} + 23.38 \text{ ft} - 7.60 \text{ ft}$$
$$= 171.24 \text{ ft}$$

Answer (C)

1. Introduction

Route surveying refers to the vertical and horizontal alignment of a fixed public or private improvement project, such as a highway or a bridge. Whether in the vertical or horizontal planes, the straight portions of the alignment are said to be *tangent*, as distinguished from those portions on a curve. In highway engineering, horizontal curves are defined as *circular*, while vertical curves are *equal-tangent parabolic*. A third type of curve, the *spiral*, is considered a special case of the circular curve and is used primarily in railroad engineering.

2. Curves

Roads, rail lines, and water courses are usually designed to be straight lines. Where a direction change is needed, a curve is used. The straight lines connected by a curve are known as *tangents* or *tangent lines*. A curve on level ground changing the direction of two tangents is known as a *horizontal curve*. Horizontal curves are usually arcs of circles.

Curves must also be used to connect roads and rail lines that change grade (slope). Those curves are called *vertical curves*. A curve that connects an upgrade tangent to a downgrade tangent is known as a *crest curve*, whereas a curve that connects a downgrade tangent to an upgrade tangent is known as a *sag curve*. Vertical curves are usually parabolic in shape.

Figure 8.1 *Sag and Crest Vertical Curves*

3. Horizontal Curves

A. Nomenclature

The elements of circular curves and their standard abbreviations are given as follows.

C the long chord—the straight distance from PC to PT

D the degree of the curve

E the external distance—the distance from V to the midpoint of the curve

I interior angle

L length of the arc—the length of the curve from PC to PT

M the middle ordinate—the distance from the curve midpoint to the midpoint of the long chord

PC   point of curvature—the place where the first tangent ends and the curve begins

PI   point of intersection

POC   any point on the curve

PT   point of tangency—the place where the curve ends and the second tangent begins

R radius of the curve

T tangent distance from V to PC or from V to PT

The following alternative designations are also sometimes used.

BC   the beginning of a curve (same as PC)

CT   a change from a curve to a tangent (same as PT)

EC   the end of a curve (same as PT)

TC   a change from a tangent to a curve (same as PC)

Δ the intersection angle (same as I)

V the vertex—the intersection of the two tangents (same as PI)

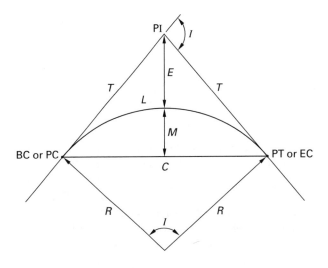

Figure 8.2 *Elements of a Circular Curve*

Equations 8.1 through 8.6 can be used to solve problems involving circular curves.

$$T = R \tan\left(\tfrac{1}{2}I\right) \tag{8.1}$$

$$E = R\left(\tan\tfrac{1}{2}I\right)\left(\tan\tfrac{1}{4}I\right) = R\left(\sec\tfrac{1}{2}I - 1\right) \tag{8.2}$$

$$M = R\left(1 - \cos\tfrac{1}{2}I\right) = \tfrac{1}{2}C\left(\tan\tfrac{1}{4}I\right) \tag{8.3}$$

$$C = 2R\left(\sin\tfrac{1}{2}I\right) = 2T\left(\cos\tfrac{1}{2}I\right) \tag{8.4}$$

$$L = R(I \text{ in radians})$$
$$= R(I \text{ in degrees})\left(\frac{2\pi}{360°}\right)$$
$$= (100 \text{ ft})\left(\frac{I}{D}\right) \tag{8.5}$$

The curvature of city streets, property boundaries, and some highways is usually specified by the radius, R. The curvature may also be specified (in degrees) by the *degree of curve*, D.

$$D = \frac{(360°)(100 \text{ ft})}{2\pi R} = \frac{5729.6°}{R} \tag{8.6}$$

In most highway work, the length of the curve is understood to be the actual arc, and the degree of the curve is the angle subtended by an arc of 100 ft. Therefore, the degree of curve can be expressed in degrees per station.

When the degree of curve is related to an arc of 100 ft, it is said to be calculated on an *arc basis*. In railroad surveys, the *chord basis* is used, and the degree of curve is the angle subtended by a chord of 100 ft. In that case, the degree of curve and radius are related by

$$\sin\left(\frac{D}{2}\right) = \frac{50 \text{ ft}}{R} \tag{8.7}$$

Where the radius is large (4° curves or smaller), the difference between the arc and chord methods is insignificant.

Stationing is continuous every 100 ft along a highway and around the curve. However, when the initial route is laid out between PIs, the curve is undefined. The route distance is measured from PI to PI. Therefore, each PI will have two stations associated with it. The *forward station* is equal to the PC station plus the tangent length. The *back station* is equal to the PT station minus the tangent length. (The PT station is equal to the PC station plus the arc length.)

B. Tangent Offsets for Circular Curves

The tangent offset shown in Fig. 8.3 can be calculated from Eq. 8.8.

$$y = R(1 - \cos\alpha) \tag{8.8}$$

$$\alpha = \arcsin\left(\frac{x}{R}\right) \tag{8.9}$$

$$x = R\sin\alpha \tag{8.10}$$

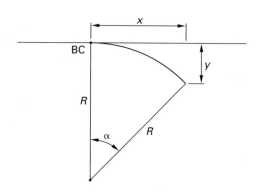

***Figure 8.3** Tangent Offset*

C. Deflection Angles

The surveyor must stake out the curve so that the road crew knows where to put the road. Stakes should be put at the PC, PT, and at all full stations. If the curve is sharp, stakes may also be required at +25, +50, and +75 stations. The *deflection angle method* is the most common method used for staking out the curve. In this method, the curve distance is usually assumed to start from $00 + 00$ at the PC.

The *deflection angle* is defined as the angle between the tangent and a chord. This is illustraded in Fig. 8.4. The deflection angles are calculated using the following theorems.

- The deflection angle between a tangent and a chord is half the subtended arc.

- The angle between two chords is half the subtended arc.

In Fig. 8.4, angle V-PC-A is a deflection angle between a tangent and a chord. The angle is

$$V\text{-PC-}A = \tfrac{1}{2}\alpha \tag{8.11}$$

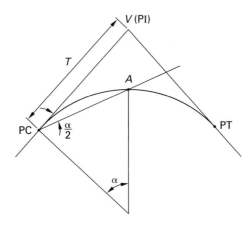

Figure 8.4 *Circular Curve Deflection Angle*

Angle α can be found from the following relationships.

$$\frac{\alpha}{360°} = \frac{\text{arc length(PC-}A)}{2\pi R} \qquad 8.12$$

$$\frac{\alpha}{I} = \frac{\text{arc length (PC-}A)}{\text{LC}} \qquad 8.13$$

The chord length PC-A is given by Eq. 8.14.

$$\text{chord length PC-}A = 2R\sin\tfrac{1}{2}\alpha \qquad 8.14$$

The entire curve can be laid out from the PC by sighting the deflection angle V-PC-A and taping the chord distance PC-A. The PC and PT can be found by solving for T and starting at V.

Example 8.1

A circular curve is to be constructed with a 225 ft radius and an interior angle of 55°. Determine where the stakes should be placed if the separation between stakes along the arc is 50 ft. Specify the first and last interior angles (subtended arcs) and the chord lengths.

Solution

The length of the curve is given by Eq. 8.5.

$$L = (225\text{ ft})(55°)\left(\frac{2\pi}{360°}\right) = 215.98\text{ ft}$$

The last stake will be 215.98 ft − 200 ft = 15.98 ft from the next to the last stake. The central angle for an arc of 50 ft is given by Eq. 8.12.

$$\alpha = \left(\frac{360°}{2\pi}\right)\left(\frac{50\text{ ft}}{225\text{ ft}}\right) = 12.732°$$

12.732° goes into 55° four times with a remainder of 4.072°. From Eq. 8.14, the required chord lengths are

$$(2)(225\text{ ft})\sin\left(\frac{12.732°}{2}\right) = 49.90\text{ ft}$$

$$(2)(225\text{ ft})\sin\left(\frac{4.072°}{2}\right) = 15.98\text{ ft}$$

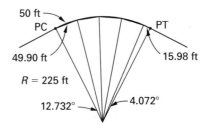

Example 8.2

An interior angle of 8.4° is specified for a horizontal curve. The PI station is $64+27.46$. Use a 2° curve and locate the PC and PT stations.

Solution

From Eq. 8.6,

$$R = \left(\frac{360°}{2°}\right)\left(\frac{100}{2\pi}\right) = 2864.79\text{ ft}$$

From Eqs. 8.1 and 8.5,

$$T = (3864.79\text{ ft})\tan\left(\frac{8.4°}{2°}\right) = 210.38\text{ ft}$$

$$L = (2864.79\text{ ft})(8.4°)\left(\frac{2\pi}{360°}\right) = 420.00\text{ ft}$$

Then the PC and PT points are located.

$$\text{PC} = (64\text{ ft}+27.46\text{ ft}) - (2\text{ ft}+10.38\text{ ft})$$
$$= 62\text{ ft}+17.08\text{ ft}$$

$$\text{PT} = (62\text{ ft}+17.08\text{ ft}) + (4\text{ ft}+20.00\text{ ft})$$
$$= 66\text{ ft}+37.08\text{ ft}$$

4. Vertical Curves

Vertical curves are used to change the grade of a highway. *Equal-tangent parabolic curves* are usually used for this purpose. A vertical sag curve connecting two grades is shown in Fig. 8.5. Since the grades are very small, the actual arc length of the curve is approximately equal to the chord length BVC-EVC.

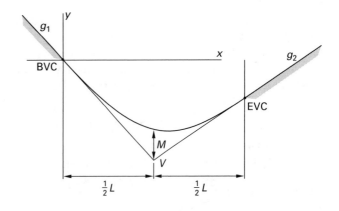

Figure 8.5 *Vertical Curve*

The following abbreviations are used.

BVC   beginning of the vertical curve

EVC   end of the vertical curve

g_1 the grade from which the stationing starts (in percent)

g_2 the grade toward which the stationing heads (in percent)

L the horizontal length of the curve (in stations)

M the middle ordinate

PVC   same as BVC

PVI   same as V

PVT   same as EVC

V the vertex—the intersection of the two tangents

A vertical parabolic curve is completely specified by the two grades and the curve length. Alternately, the *rate of grade change* per station can be used in place of the curve length.

$$r = \frac{g_2 - g_1}{L} \qquad 8.15$$

Equation 8.16 defines an equal-tangent parabolic curve. x is measured in stations beyond BVC. Elevation (elev) is measured in feet, with the same reference point used to measure all elevations.

$$\text{elev}_x = \left(\frac{r}{2}\right) x^2 + g_1 x + \text{elev}_{\text{BVC}} \qquad 8.16$$

The maximum or minimum elevation will occur at the *turning point*. The turning point is not located directly above or below V, but is found at

$$x = \frac{-g_1}{r} \quad \text{[in stations]} \qquad 8.17$$

The *middle ordinate* distance is

$$M = \frac{|(g_1 - g_2)|(L)}{8} \qquad 8.18$$

Example 8.3

A vertical crest curve with a length of 400 ft is to connect grades of +1.0% and −1.75%. The vertex is located at station 35+00, and it has an elevation of 549.20 ft. What are the elevations of the BVC and EVC, and at all full stations on the curve?

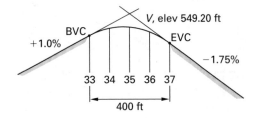

Solution

The elevation at BVC is 549.20 ft − (1)(2) = 547.20 ft.

The elevation at EVC is 549.20 ft − (1.75)(2) = 545.70 ft.

$$r = \frac{-1.75\% - 1}{4} = -0.6875\%$$

$$\tfrac{1}{2}r = -0.3438\%$$

The equation of the curve is

$$y = -0.3438x^2 + x + 547.20 \text{ ft}$$

At sta 34, $x = 34 - 33 = 1$. So,

$$y_{34} = (-0.3438)(1)^2 + 1 + 547.20 \text{ ft} = 547.86 \text{ ft}$$

Similarly,

$$y_{35} = (-0.3438\%)(2)^2 + 2 + 547.20 \text{ ft} = 547.82 \text{ ft}$$
$$y_{36} = (-0.3438\%)(3)^2 + 3 + 547.20 \text{ ft} = 547.11 \text{ ft}$$

5. Asymmetrical Vertical Curves or Vertical Curves Having Unequal Tangents

Occasionally, a curve must be a specific elevation at a certain station and the grades of the grade lines cannot be changed. In this situation, a curve with unequal tangents must be used. There are mathematical formulas to compute the elements of the asymmetrical curve, however the most practical way to deal with such a curve is to treat it as two curves, joined at a point of compound vertical curve (PCVC).

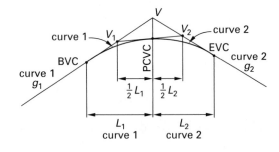

Figure 8.6 Asymmetrical Vertical Curve

The properties of the asymmetrical curve in Fig. 8.6 are

- V_1 and V_2 are the PIs for curves 1 and 2   8.19
- grade line $V_1 - V_2$ is tangent to the curve at the PCVC   8.20
- length BVC $- V_1 = V_1 - V$ 8.21
- length $V - V_2 = V_2 - $ EVC   8.22

- elevation V_1 is computed midway between BVC and V ⎫ 8.23
- elevation V_2 is computed midway between V and EVC ⎫ 8.24
- length $V_1 - V_2 = (V_1 - V) + (V - V_2)$ ⎫ 8.25
- grade $V_1 - \text{PCVC} = g_2$ for curve 1 ⎫ 8.26
- grade $\text{PCVC} - V_2 = g_1$ for curve 2 ⎫ 8.27

The asymmetrical vertical curve can be treated as two separate curves, and the equations for the symmetrical curves can then be applied to each.

6. Spiral Curves

Spiral curves are used to produce a gradual transition from horizontal tangents to horizontal circular curves in roadways and railroads. A spiral curve is a curve of gradually changing radius and gradually increasing or decreasing degree of curvature. Figure 8.7 illustrates a spiral curve, showing the *tangent to spiral* (TS) point, the *length of spiral* (LS), and the *spiral to curve* (SC) point.

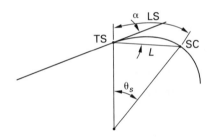

Figure 8.7 *Simple Spiral Curve*

The *degree of spiral* changes continuously, so the average degree of spiral is used to calculate the length. (θ_s is in degrees in Eq. 8.28.)

$$\text{LS} = \frac{(100)\theta_s}{\dfrac{D}{2}} \qquad 8.28$$

Deflection angles for setting out points on the spiral curve can be calculated from Eq. 8.29.

$$\alpha = \left(\frac{\theta_s}{3}\right)\left(\frac{L}{\text{LS}}\right)^2 \qquad 8.29$$

When used in transportation facilities, experiments suggest that lengths of spiral curves should be based on the speed of traffic entering the curve and the radius of the curve being approached.

$$\text{LS} = \frac{(1.6)(\text{mph})^3}{R \text{ in feet}} \qquad 8.30$$

7. Estimating Earthwork Volumes

The soil between two stations is known as a soil *prismoid* or *prism*. The prismoid volume must be calculated in order to estimate haulage requirements. Such volume is generally expressed in cubic yards. There are two methods of calculating the prismoid volume: the average end area method and the prismoidal formula.

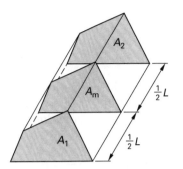

Figure 8.8 *Soil Prismoid*

In the *average end area method*, the volume is calculated by averaging the two end areas and multiplying by the prism length. This method disregards the slopes and orientations of the ends and sides, but is sufficiently accurate for earthwork calculations. When the end area is complex, it may be necessary to use a planimeter or to plot the area on fine grid paper and simply count the squares. The average end area method usually overestimates the actual soils volume and thus favors the contractor in earthwork costing. In Eq. 8.31, L is in feet, A is in square feet, and V is in cubic yards.

$$V = \frac{L(A_1 + A_2)}{(2)\left(27\ \dfrac{\text{ft}^3}{\text{yd}^3}\right)} \qquad 8.31$$

The precision obtained from the average end area method is generally sufficient unless one of the end areas is very small or zero. In that case, the volume should be computed as a pyramid or truncated pyramid.

$$V_{\text{pyramid}} = \frac{L A_{\text{base}}}{(3)\left(27\ \dfrac{\text{ft}^3}{\text{yd}^3}\right)} \qquad 8.32$$

The *prismoidal formula* is better suited when the two end areas differ greatly or when the ground surface is irregular. It generally will produce a smaller volume than the average end area method and thus favors the owner-developer in earthwork costing. The prismoidal formula uses the area, A_m, midway between the two end sections. In the absence of actual observed measurements, the dimensions of the middle area can be found by averaging the similar dimensions of the two end areas. (The middle area is not found by averaging the two end areas.)

$$V = \left[\frac{L}{(6)\left(27\ \dfrac{\text{ft}^3}{\text{yd}^3}\right)} \right] (A_1 + 4A_m + A_2) \qquad 8.33$$

Practice Problems

1. What is the length of storm pipe that is constructed between a point 50.00 ft left of sta 3+50 and 50.00 ft right of sta 5+00? The radius of the centerline is 600.00 ft.

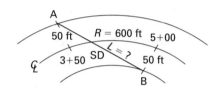

(A) 150.00 ft

(B) 179.52 ft

(C) 600.00 ft

(D) $2\pi RD/360°$

2. A circular curve whose radius is 800 ft is located between two tangents bearing N 65°30′15″ E and N 73°45′35″ E. Solve for length, semitangent, and long chord.

3. Using the data from Prob. 2, what is the difference (in feet) between the mid-ordinate and the external distance?

(A) 0.001 ft

(B) 0.0054 ft

(C) 0.0345 ft

(D) 1.026 ft

4. Given the following curve data, if the instrument is set on station 16+25 and the sighting station is 15+75, what is the deflection angle to be turned to set station 17+25?

$$R = 1000.00$$
$$\text{BC sta} = 15+36.25$$
$$\text{EC sta} = 18+40.69$$

(A) 01°06′36″

(B) 04°17′15″

(C) 05°24′26″

(D) 08°35′40″

5. A curve, defined by arc, has a degree of curve 03°49′11″. What is the length of curve if the central angle is 10°08′06″?

(A) 132.67 ft

(B) 264.98 ft

(C) 265.33 ft

(D) 530.67 ft

6. If the PI elevation of a sag vertical curve is 453.00 ft at sta 46+50, $g_1 = -1.5\%$, $g_2 = +2.25\%$, and the curve length is 600.00 ft, compute the elevations on the curve for each half station.

7. A 2% grade line intersects a 4% grade line. The elevation at sta 15+00 on the 2% line is 350.00 ft, and the elevation of station 25+00 on the 4% line is 356.00 ft. What are the station and elevation of the intersection?

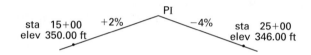

(A) 19+00 at 358.00 ft

(B) 20+00 at 360.00 ft

(C) 20+50 at 351.00 ft

(D) 21+00 at 362.00 ft

8. Place a 400 ft vertical curve between the tangents in Prob. 7. What are the station and elevation of the high point on the curve?

(A) 19+66.67 at 359.00 ft

(B) 20+00.00 at 359.34 ft

(C) 20+33.33 at 359.34 ft

(D) 21+00.00 at 359.00 ft

Solutions

1. Construct a triangle using A, B, and the radius point of the 600.00 radius curve.

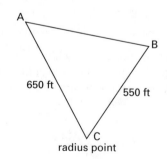

Since A and B are radial to the centerline, angle ABC is the central angle (Δ) between 3+50 and 5+00, a distance of 150.00 ft on the centerline. Solve for the Δ.

$$L = R \left(\frac{2\pi}{360} \right) \Delta$$

$$\Delta = \frac{L(360°)}{R(2\pi)} = \frac{(150 \text{ ft})(360°)}{(600 \text{ ft})(2\pi)} = 14.3239°$$
$$= 14°19'26''$$

Solve the triangle for side AB using the law of cosines.

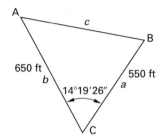

$$c^2 = a^2 + b^2 - 2ab \cos C$$

$$c = \sqrt{a^2 + b^2 - 2ab \cos C}$$
$$= \sqrt{\begin{array}{c}(550 \text{ ft})^2 + (650 \text{ ft})^2 \\ - (2)(550 \text{ ft})(650 \text{ ft})(\cos 14°19'26'')\end{array}}$$
$$= \sqrt{32{,}227.44 \text{ ft}^2}$$
$$= 179.52 \text{ ft}$$

Answer (B)

2. $\qquad \Delta = 73°45'35'' - 65°30'15'' = 08°15'20''$

$$L = \frac{2\pi R \Delta}{360°} = 115.27 \text{ ft}$$

$$\text{semitangent } (T) = R \tan \left(\frac{\Delta}{2} \right) = 57.73 \text{ ft}$$

$$L = 2R \sin \left(\frac{\Delta}{2} \right) = 115.17 \text{ ft}$$

3. $\qquad \text{external} = R \left[\left(\frac{1}{\cos \dfrac{\Delta}{2}} \right) - 1 \right] = 2.0806 \text{ ft}$

$$\text{mid-ordinate} = R \left[1 - \cos \left(\frac{\Delta}{2} \right) \right] = 2.0752 \text{ ft}$$

$$2.0806 \text{ ft} - 2.0752 \text{ ft} = 0.0054 \text{ ft}$$

Answer (B)

4. Compute the length of curve.

$$L = (18 \text{ ft} + 40.69 \text{ ft}) - (15 \text{ ft} + 36.25 \text{ ft}) = 304.44 \text{ ft}$$

Compute the central angle.

$$L = 2\pi R \left(\frac{\Delta}{360°} \right)$$

$$\Delta = \frac{L(360°)}{2\pi R} = 17.4431°$$

Compute $\Delta/2$.

$$\frac{17.4431°}{2} = 8.7216° = 08°43'18''$$

This is full deflection.

Compute the deflection per foot.

$$\frac{D}{\text{ft}} = \frac{\dfrac{\Delta}{2}}{L} = \frac{8.7216°}{304.44 \text{ ft}} = 0.02865°/\text{ft}$$

Compute deflections for sta 15+75 and sta 17+25.

$$\left[\begin{array}{c} (15 \text{ ft} + 75 \text{ ft}) \\ - (15 \text{ ft} + 36.25 \text{ ft}) \end{array} \right] \left(\frac{0.02865°}{\text{ft}} \right) = 1.11011°$$

$$\left[\begin{array}{c} (17 \text{ ft} + 25 \text{ ft}) \\ - (15 \text{ ft} + 36.25 \text{ ft}) \end{array} \right] \left(\frac{0.02865°}{\text{ft}} \right) = 5.40729°$$

Compute difference between deflections.

$$5.40729° - 1.11011° = 4.29718°$$
$$= 04°17'15''$$

Therefore, site on sta 16+25, sight sta 15+75, and turn the deflection angle 04°17'15'' to sta 17+25.

Answer (B)

5. There are several solutions to this problem.

Solution one: By definition, the arc length for a degree of curve of 03°49'11'' is 100.00 ft (x is the length of curve). To find the total length, set up a ratio.

$$\frac{03°49'11''}{100 \text{ ft}} = \frac{10°08'06''}{x}$$

Convert the angles and solve for x.

$$x = 265.33 \text{ ft}$$

Solution two: Compute the radius of a curve having a Δ of $03°49'11''$ and a length of 100.00 ft.

$$L = 2\pi R \left(\frac{\Delta}{360°} \right)$$

$$R = \frac{(360°)L}{2\pi D} = 1500.00 \text{ ft}$$

Compute the length of curve for a Δ of $10°08'06''$ and a radius of 1500.00 ft.

$$L = 2\pi R \left(\frac{\Delta}{360°} \right) = 265.33 \text{ ft}$$

Answer (C)

6. Compute the BVC and EVC grades.

$$\begin{array}{l}(\text{sta 43+50 BVC} - 453.00 \text{ ft}) \\ + (1.5\%)(3 \text{ sta}) \end{array} = 457.50$$

$$\begin{array}{l}(\text{sta 49+50 EVC} - 453.00 \text{ ft}) \\ + (2.25\%)(3 \text{ sta}) \end{array} = 459.75$$

Compute M.

$$M = \frac{|(g_1 - g_2)|(L)}{8} = \frac{|(-1.50\% - 2.25\%)|(6 \text{ sta})}{8}$$

$$= \frac{|(-3.75\%)|(6 \text{ sta})}{8}$$

$$= +2.81 \text{ sta}$$

Compute the grade line elevations and the offsets to the curve from the grade line. Add the offsets to the grade line elevations for curve grades.

$$d = M \left(\frac{D}{\frac{L}{2}} \right)^2 \qquad \text{[column 5 formula]}$$

(1) sta	(2) distance to BVC = D	(3) col 2g 1.5%	(4) BVC elev − col 3 = gr line elev	(5) offset from gr line to curve	(6) curve elev col 4 + col 5 (d)
43+50	0.00	0.00	457.50	0.00	457.50
44+00	0.50	0.75	456.75	0.08	456.83
44+50	1.00	1.50	456.00	0.31	456.31
45+00	1.50	2.25	455.25	0.70	455.95
45+50	2.00	3.00	454.50	1.25	455.75
46+00	2.50	3.75	453.75	1.95	455.70
46+50	3.00	4.50	453.00	2.81	455.81
47+00	3.50	5.25	452.25	3.82	456.07
47+50	4.00	6.00	451.50	4.99	456.50
48+00	4.50	6.75	450.75	6.32	457.07
48+50	5.00	7.50	450.00	7.80	457.80
49+00	5.50	8.25	449.25	9.45	458.70
49+50	6.00	9.00	448.50	11.25	459.75

7. Set the diagram up as follows.

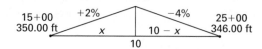

The difference in elevation in 10 stations is 350.00 ft − 346.00 ft. Using the grade lines as a means to show the difference in elevation yields

$$2x - (4)(10 - x) = -4.0 \text{ ft}$$

$$2x - 40 + 4x = -4.0 \text{ ft}$$

$$6x = 36$$

$$x = 6 \text{ sta}$$

The station of the PI is $15+00 + 6.0 \text{ sta} = 21+00$.

The elevation at 21+00 is 350.00 ft + (2%)(6 sta) = 362.00.

Answer (D)

8. Compute the BVC and EVC stations.

$$\text{BVC} = 21+00 - 2 \text{ sta} = 19+00$$

$$\text{EVC} = 21+00 + 2 \text{ sta} = 23+00$$

Compute the BVC elevation.

$$\text{BVC} = 362.00 \text{ ft} - (2\%)(2 \text{ sta}) = 358.00 \text{ ft}$$

Compute the station of the high point.

$$D_o = \frac{LG}{g_2 - g_1} = \frac{(4 \text{ sta})(2\%)}{6\%} = \frac{8}{6} = 1.3333 \text{ sta}$$

$$19+00 + 1.3333 = 20+33.33 \quad \text{[sta of high point]}$$

Compute the grade line elevation at 20+33.33.

$$358.00 \text{ ft} + (2)(1.3333) = 360.67 \text{ ft}$$

Compute M.

$$M = \frac{|(g_1 - g_2)|(L)}{8} = \frac{(6)(4)}{8} = 3.00$$

Compute offset from grade line to curve d.

$$d = M \left(\frac{D}{\frac{L}{2}} \right)^2 = \left(\frac{1.3333}{2} \right)^2 = 1.33 \text{ ft}$$

Compute curve elevation.

$$360.67 - 1.33 = 359.34 \text{ ft}$$

Answer (C)

Survey Staking 9

1. Introduction

Control of alignment and grade during construction is established through the use of stakes placed by either the owner or the contractor on the project. The stakes are placed and coded such that earthwork, trenching, and subgrade can be constructed to the designed lines and grades. The coding, or markings, on the stakes is derived from both the design parameters and the existing field parameters.

2. Stakes

Wood stakes used by surveyors are commonly referred to as *construction stakes*. Depending on their use, construction stakes are also called *alignment stakes*, *offset stakes*, *grade stakes*, and *slope stakes*. Stakes come in many sizes, including 1 in $\times$ 2 in $\times$ 18 in markers, $5/16$ in $\times$ $1^1/2$ in $\times$ 24 in half-lath stakes, and various lengths of 2 in $\times$ 2 in hubs and 1 in $\times$ 1 in guineas. Hubs and guineas are usually 12 or 18 in long and occasionally may measure 24 in in length.

The front (the side facing the construction) and back (the side facing away from construction) of the stakes are marked with keel, carpenter's crayon, or permanent ink markers. Stakes are read from top to bottom. Distances are measured in feet, and a maximum precision of 0.1 ft is considered standard for earthwork. A precision of 0.01 ft is standard for locating positions and elevations of installed structures. Pin flags or full-length lagging may be used to provide further identification and visibility.

Front-of-stake markings include header information (e.g., *reference point slope stake*, RPSS, offset distance) and cluster information (e.g., horizontal and vertical measurements, slope ratio). The header is separated from the first cluster by a double horizontal line. Multiple clusters are separated by single horizontal lines. All clusters are measured in the same direction from the same point.

There is great variation in stake marking conventions. The conventions used in this book are common but not universal practice. Words like "from," "above," and "below" are understood and are seldom actually written

on a stake. For example, "2.0 FC" and "4.3 FG" mean "2.0 ft from the face of the curb" and "4.3 ft above the finished grade," respectively.

The back of a stake is usually used to record the station or other literal locating information (e.g., "at ramp"). Actual elevations, when included, are marked on the side of the stake, not on the front or back. Other information identifying the survey may also be included on the back.

In its simplest use, a stake both locates and identifies a specific point. A distance referenced on a stake is measured from the natural ground at the point where the stake is driven. Another less frequently used system involves measuring from a tack or a reference mark, such as a crow's foot, on the stake itself. A *crow's foot* consists of a horizontal line drawn on the stake with a vertical arrow pointing to it. When used, a crow's foot should be drawn on the side, not on the front or back of the stake.

Many times, however, measurements are taken from a *hub stake* (also known as a *ground stake*, *reference point stake*, or RP) driven essentially flush with the ground. No markings are made on ground stakes. The term *blue top* can mean a ground stake that is pounded to grade or, more generally, a stake on which grading information is noted.

Hubs and other ground stakes are located, identified, and protected by *witness stakes* (also known as *guard stakes*). A witness stake documents a ground stake but does not itself locate a specific point, and it may be driven at an angle with its top over the flush-driven stake.

Alignment stakes marked with the station indicate the centerline alignment of roads and highways. They are placed every 50 or 100 ft station along highway tangents with uniform grade, and every 25 or 50 ft station for horizontal and vertical curves.

Offset stakes used to mark excavations or roads for paving are offset from the actual edge to protect the stakes from construction equipment. The offset distance is circled on the stake and is separated from the subsequent data by a double line. For close-in work, the offset distance is often standardized at 2 ft. For highway work, offset stakes (measured from the alignment centerline) can be set 25, 50, or 100 ft from the centerline on both sides. Unless a separate ground stake (hub) is used, distances (e.g., cuts, fills, and distances to centerline) marked on an offset stake refer to the point of insertion, not to the imaginary point located the offset distance away.

Figure 9.1 illustrates how a construction stake would be marked to identify a trench for a storm drain.

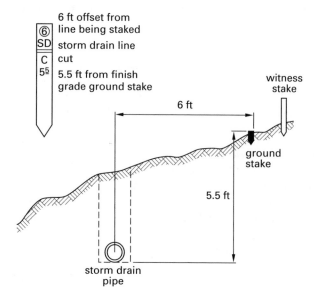

Figure 9.1 *Construction Stake for Storm Drain*

Slope stakes indicate *grade points* (i.e., points where cuts and fills begin and the planned side slopes intersect the natural ground surface). Slope stakes are marked SS to indicate their purpose. (It is unusual for a slope stake to be set at the actual points where a cut or fill begins. The typical position is on a 10 ft offset.)

In addition to indicating the grade point, the front of slope stakes are marked to indicate the nature of the earthwork (i.e., C for cut and F for fill), the offset, the type of line being staked, the distance from the roadbed centerline or other stationed control line, the slope (horizontal:vertical) to finished grade, and the elevation difference between the grade point and the finished grade. Distances from the centerline can be marked R or L to indicate whether the stake points are to the right or left of the centerline when looking up-station. The station, reckoned along the centerline, is marked on the back of the stake.

For example, the front-of-stake markings "C 4.2 FG @ 38.4L CL 2:1" would be interpreted as "cut with 2:1 slope is required; stake set point is 4.2 ft above finished grade; stake is 38.4 ft to the left of the centerline of roadway." The use of FG is optional.

Slope stakes can be driven vertically or at an angle, depending on convention. When driven at an angle, slope stakes slant outward (point away from the earthwork) when fill is required and inward (point into the earthwork) when a cut is required. The stakes are set with the broad (front) face toward the construction.

Ground stakes (hubs) without offsets are not used with fill stakes, as a ground stake would soon be covered. For shallow and moderate fills, a fill stake alone is used. A crow's foot may be marked on the stake to indicate the approximate finished grade, or the top of the stake may be set to coincide with the approximate finished grade. For deeper fills, offset stakes are required. Fill stakes are sometimes ripped diagonally lengthwise to give them a characteristic shape.

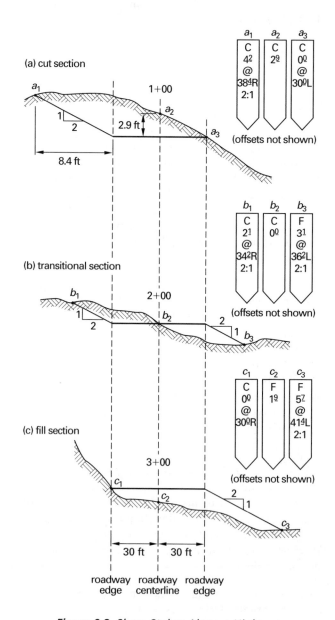

Figure 9.2 *Slope Stakes Along a Highway*

Not all earthwork produces a straight slope. Some earthwork results in a smooth-curved ground surface between two lateral points. The points where the grading begins and ends are known as the *begin slope rounding* (BSR) and *end smooth rounding* (ESR) points, rather than the analogous grade point.

A line of stakes at adjacent grade or ESR points is known as a *daylight line.* Therefore, the term *daylight stake* can be used when referring to the stakes marking the grade/ESR points. (A mole or worm following the finished slope would emerge from the ground and see daylight at the grade/ESR point, hence the term *daylight point.*)

Figure 9.2 illustrates the use of slope stakes along three adjacent sections of a proposed highway.

Example 9.1

During your review of a surveyor's field work along a highway, you encounter a ground stake and its accompanying illustrated witness stake. What is the elevation of the toe of the slope documented by the witness stake?

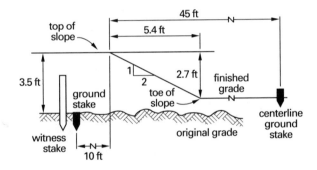

Solution

The witness stake indicates that the ground stake is a reference point slope stake located 45.0 ft to the right (when looking up-station) of the centerline of the highway. The elevation of the ground stake, marked on the side of the witness stake, is 87.6 ft. The earthwork starts out with a 3.5 ft fill. Then, the earth is cut away at a 2:1 slope until the finished grade is reached, 2.7 ft below.

The elevation of the toe of the slope is the same as the elevation of the finished grade.

$$\text{slope toe elevation} = \text{elevation of ground stake}$$
$$+ \text{ fill} - \text{cut}$$
$$= 87.6 \text{ ft} + 3.5 \text{ ft} - 2.7 \text{ ft}$$
$$= 88.4 \text{ ft}$$

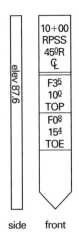

3. Establishing Slope Stake Markings

The markings on a construction stake are determined from an initial survey of the natural ground surface. This survey required two individuals—one (the leveler or instrumentperson) to work the instrument, the other (the rodperson) to hold the leveling rod. The elevation of the instrument (or, alternatively, the elevation of the ground at the instrument location and the height of the instrument above the ground) must be known if actual elevations are to be marked on a stake. (*Height of the instrument,* or HI, means "elevation of the instrument" in standard practice; it does not mean the instrument distance above the ground.)

$$\text{HI} = \text{ground elevation}$$
$$+ \text{ instrument altitude above ground} \quad 9.1$$

The *rod reading for ground* (commonly referred to as the *ground rod*) is the sighting made through a leveling instrument at the rod held vertically at the grade point. The *grade rod,* short for *rod reading for grade,* is the reading that would be observed on an imaginary rod held on the finished grade elevation. The grade rod (reading) is calculated from the planned grade elevation and the height of the instrument.

$$\text{grade rod} = \text{HI} - \text{grade elevation} \quad 9.2$$

The distance marked on a construction stake (i.e., the cut or fill) is the difference in natural and finished grade elevations and is easily calculated from the ground and grade rods (readings). The cut or fill stake marking is the difference between grade and ground rods. The actual steps taken to calculate the stake marking depend on whether the earthwork is a cut or a fill and whether the instrument is above or below the finished grade. Drawing a diagram will clarify the algebraic steps and prevent sign errors.

The distance from the grade point to the centerline of the roadbed or other control is also written on the stake.

If w is the width of the finished surface (e.g., a roadbed), s is the side slope ratio (horizontal:vertical), and h is the cut or fill at the grade point, then the horizontal distance, d, from the grade point to the centerline stake of the finished surface is

$$d = \tfrac{1}{2}w + hs \qquad 9.3$$

In the field, the actual stake location is found by trial and error.

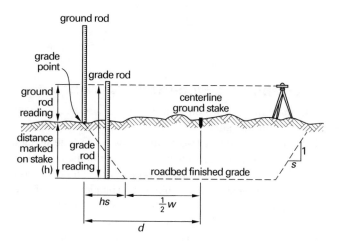

Figure 9.3 *Determining Stake Location*

Practice Problems

1. Two dimensions of a building are shown on an architect's plan as (A) 67' 9¾" and (B) 126' 6⅝". Convert the dimensions to feet and decimals of a foot.

Use the following information for Probs. 2 through 5.

The centerline of B Street is on a curve from sta 17+36.42 BC to sta 21+91.13 EC.

The curve data for this segment is

$$\text{radius } (R) = 600.00 \text{ ft}$$
$$\text{central angle } (\Delta) = 43°25'19''$$
$$\text{length } (l) = 454.71 \text{ ft}$$

The designed curb line is 40.00 ft left of centerline and 44.00 ft right of centerline. The curve is oriented to the right as you travel up station (see illustration below).

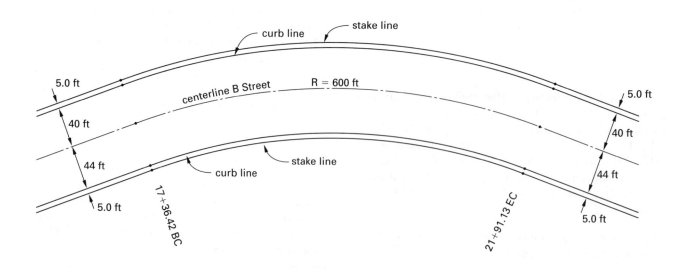

2. Compute the deflection angles needed to lay out the centerline on half-stations, starting at the BC.

3. Compute the outside chord lengths needed to stake the left curb on a 5.00 ft offset outside.

4. Compute the inside chord lengths needed to stake the right curb on a 5.00 ft offset inside.

5. The designed top of curb grade on the left, at sta 17+50, EVC, is 611.45 ft and at sta 20+00, BVC, is 614.58 ft. The HI of the instrument is 620.75 ft. If the ground rod on the offset stake at sta 19+00 is 6.10 ft, what is the cut or fill to the top of curb from that stake?

   (A) F 1.32

   (B) C 1.32

   (C) F 3.20

   (D) C 3.20

6. The centerline grade at sta 25+00 is 500.25. What is the grade at the left edge of shoulder?

   (A) 499.73 ft

   (B) 499.75 ft

   (C) 500.76 ft

   (D) 500.78 ft

7. The elevation of the right slope stake is 465.3. What is the distance from the slope stake to the centerline?

   (A) 69.4 ft

   (B) 111.4 ft

   (C) 120.6 ft

   (D) 129.4 ft

8. How would the front of the slope stake in Prob. 7 be marked?

Use the following illustration for Probs. 6 through 8.

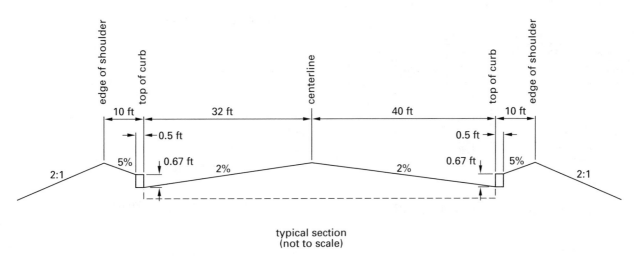

typical section
(not to scale)

sta 25+00

9. A sewer is to be constructed from an existing manhole, #1, designed invert elevation 120.79, to a proposed manhole, #2, designed invert elevation 117.91. The existing manhole is at sta 16+75, and the proposed manhole is at sta 20+25. Stakes are to be set at half-stations and are to be 10.00 ft above the flow line of the sewer. What elevation will the stake at sta 18+00 be set?

(A) 119.76 ft

(B) 121.82 ft

(C) 129.76 ft

(D) 131.82 ft

10. Slope stakes indicate which of the following?

(A) a line

(B) a grade point

(C) an offset

(D) all of the above

11. What type of stakes are required for deeper fills?

(A) offset stakes

(B) ground stakes

(C) alignment stakes

(D) slope stakes

The following information applies to Probs. 12 through 14.

sta	BS	HI	FS	elev	grade
BM #1	2.12			102.23	
0+00			5.60		88.00
0+75			8.20		91.15
TP #1	1.67		11.32		
1+75			2.10		
2+00			4.20		

12. What is the stake elevation at sta 2+00?

(A) 80.50 ft

(B) 81.00 ft

(C) 90.00 ft

(D) 90.50 ft

13. What is the grade at sta 2+00?

(A) 84.60 ft

(B) 87.40 ft

(C) 87.60 ft

(D) 96.40 ft

14. What is the fill at sta 1+75?

(A) 2.00

(B) 2.50

(C) 2.75

(D) 3.50

Solutions

1. Convert the fractions to decimals of an inch.

(A) $\frac{3}{4}$ in $= 0.75$ in

(B) $\frac{5}{8}$ in $= 0.625$ in

Add the decimal to the whole inch.

(A) 9 in $+ 0.75$ in $= 9.75$ in

(B) 6 in $+ 0.625$ in $= 6.625$ in

Divide the inch dimension by 12 to convert to decimals of a foot.

(A) $\dfrac{9.75 \text{ in}}{12} = 0.8125$ ft

(B) $\dfrac{6.625 \text{ in}}{12} = 0.552$ ft

Add the decimals of the foot to the whole foot dimension.

(A) 67 ft $+ 0.812$ ft $= 67.812$ ft

(B) 126 ft $+ 0.552$ ft $= 126.552$ ft

2. Compute the deflection per foot.

$$\frac{\text{def}}{\text{ft}} = \frac{\frac{\Delta}{2}}{L} = \frac{21°42'40''}{454.71 \text{ ft}} = \frac{21.7111°}{454.71 \text{ ft}}$$
$$= 0.047747°/\text{ft}$$

The deflection angle of any station of the curve is the deflection per foot multiplied by the distance from the station to the BC.

Compute the deflection angles using the following table.

1 sta	2 distance to BC	3 deflection per ft	4 deflection angle = (col 2)(col 3)	5 deflection angle convert to
17+36.42 BC	0	0.047747°	0°	0°
17+50	13.58	0.047747°	0.648404°	0°38′54″
18+00	63.58	0.047747°	3.035754°	3°02′09″
18+50	113.58	0.047747°	5.423104°	5°25′23″
19+00	163.58	0.047747°	7.810278°	7°48′37″
19+50	213.58	0.047747°	10.197778°	10°11′52″
20+00	263.58	0.047747°	12.585000°	12°35′06″
20+50	313.58	0.047747°	14.972222°	14°58′21″
21+00	363.58	0.047747°	17.359854°	17°21′35″
21+50	413.58	0.047747°	19.746944°	19°44′50″
21+91.13 EC	454.71	0.047747°	21.711111°	21°42′40″

Note that the deflection angle at the BC is zero, and at the EC it is half of the central angle ($\Delta/2$) of the curve.

3. The outside radius equals the centerline radius plus the distance from centerline to the curb plus the offset.

outside radius − 600 ft + 40 ft + 5 ft = 645.00 ft

The chord length equals the sine of the deflection between station times twice the radius.

$$\text{chord} = 2R\sin\left(\frac{\Delta}{2}\right)$$

Compute the chords using the following table.

1 station to station	2 deflection between stations	3 chord = $2R\sin$ (deflection) $2R = 1290.00$
17+36.42 − 17+50	0°38′54″	14.60
50′ chord	02°23′14″	53.73
21+50 − 21+91.13	01°57′50″	44.21

4. inside radius = 600.00 ft − 44.00 ft − 5.00 ft

        = 551.00 ft

1 station to station	2 deflection between stations	3 chord = $2R\sin$ (deflection) $2R = 1102.00$
17+36.42 − 17+50	0°38′54″	12.47 ft
50′ chord	02°23′14″	45.90 ft
21+50 − 21+91.13	01°57′50″	37.77 ft

5. Determine the gradient or rate of slope between stations 17+50 and 20+00.

$$S = \frac{\text{difference in elevation}}{\text{distance}}$$

$$= \frac{614.58 \text{ ft} - 611.45 \text{ ft}}{20{+}00 - 17{+}50} = \frac{3.13 \text{ ft}}{250 \text{ ft}}$$

$$= 0.0125 \quad (+1.25\%)$$

Determine the design top of the curb grade at sta 19+00 left.

$$\text{difference in grade} = (\text{distance between stations})s$$

$$= (19{+}00 - 17{+}50)(0.0125)$$

$$= (150 \text{ ft})(0.0125)$$

$$= +1.88 \text{ ft}$$

$$\text{grade at } 19{+}00 = 611.45 \text{ ft} + 1.88 \text{ ft}$$

$$= 613.33 \text{ ft}$$

Determine the grade rod at sta 19+00.

$$\text{grade rod} = \text{HI} - \text{designed grade}$$

$$= 620.75 \text{ ft} - 613.33 \text{ ft}$$

$$= 7.42 \text{ ft}$$

Determine the cut or fill at sta 19+00.

$$\text{cut or fill} = \text{grade rod} - \text{ground rod}$$

$$= 7.42 \text{ ft} - 6.10 \text{ ft}$$

$$= 1.32 \text{ ft}$$

Since the grade rod is greater than the ground rod, this is a cut situation.

Answer (B)

6. Using the cross fall gradients, compute the difference in grade from the centerline to the shoulder.

The centerline to the bottom of curb is

$$-2\% = 0.02s$$

$$(0.02)(32 \text{ ft}) = -0.64 \text{ ft}$$

$$\text{curb height} = +6.67 \text{ bottom to top of curb}$$

The top of curb to shoulder is

$$(+5\%)(9.5 \text{ ft}) = (0.05)(9.5 \text{ ft}) = +0.048 \text{ ft}$$

$$\text{difference} = -0.64 \text{ ft} + 0.67 \text{ ft} + 0.48 \text{ ft}$$

$$= +0.51 \text{ ft}$$

$$\text{shoulder grade} = \text{centerline grade} + \text{difference}$$

$$= 500.25 \text{ ft} + 0.51 \text{ ft}$$

$$= 500.76 \text{ ft}$$

Answer (C)

7. Compute the grade at the right edge of shoulder.

$$\text{centerline grade} = 500.25 \text{ ft}$$

$$\text{bottom of curb} = 500.25 \text{ ft} - [(40.00 \text{ ft})(0.02\%)]$$

$$= 500.25 \text{ ft} - 0.80 \text{ ft}$$

$$= 499.45 \text{ ft}$$

$$\text{top of curb} = 499.45 \text{ ft} + 0.67 \text{ ft}$$

$$= 500.12 \text{ ft}$$

$$\text{shoulder grade} = 500.12 \text{ ft} + [(9.50 \text{ ft})(0.05\%)]$$

$$= 500.12 \text{ ft} + 0.48 \text{ ft}$$

$$= 500.60 \text{ ft}$$

Compute the fill at the slope stake.

$$\text{fill} = \text{shoulder grade} - \text{slope stake elevation}$$

$$= 500.60 \text{ ft} - 465.3 \text{ ft}$$

$$= 35.30 \text{ ft}$$

Compute the distance from the slope stake to the shoulder.

$$\text{distance} = (\text{fill})(2) \quad [\text{2:1 slope}]$$

$$= (35.30 \text{ ft})(2)$$

$$= 70.60 \text{ ft}$$

Compute the distance from the stake to the centerline.

$$\text{distance to centerline} = \text{distance to shoulder}$$

$$+ \text{roadway width (right)}$$

$$= 70.60 \text{ ft} + 50 \text{ ft}$$

$$= 120.60 \text{ ft}$$

Answer (C)

8. There are several options available regarding how to mark the stake.

$$\text{F } 35^3 \text{ @ 2:1} \qquad [\text{no distance}]$$

$$\text{F } 35^3 \text{ @ } 70^6 \text{ ES 2:1} \quad [\text{distance to shoulder}]$$

$$\text{F } 35^3 \text{ @ } 120^6 \text{ L 2:1} \quad [\text{distance to centerline}]$$

9. Compute the slope of the pipe.

$$\text{slope} = \frac{\text{difference in elevation}}{\text{distance between manholes}}$$

$$= \frac{120.79 \text{ ft} - 117.91 \text{ ft}}{120{+}25 - 16{+}75}$$

$$= \frac{2.88 \text{ ft}}{350 \text{ ft}}$$

$$= 0.0082 \quad (0.82\%)$$

Compute the pipe elevation at sta 18+00.

$$\begin{aligned}
\text{pipe elevation} &= \text{elevation MH \#1} \\
&\quad - (\text{distance})(\text{slope}) \\
&= 120.79 \text{ ft} - (18{+}00 - 16{+}75)(0.0082 \text{ ft}) \\
&= 120.79 \text{ ft} - (125.00)(0.0082 \text{ ft}) \\
&= 120.79 \text{ ft} - 1.03 \text{ ft} \\
&= 119.76 \text{ ft}
\end{aligned}$$

Check: 18+00 to MH #2 @ 20+25

$$\text{distance to MH \#2} = 20{+}25 - 18{+}00 = 225 \text{ ft}$$

$$\text{difference in elevation} = (225 \text{ ft})(0.082) = 1.85 \text{ ft}$$

$$\begin{aligned}
\begin{array}{c}\text{pipe elevation} \\ \text{at MH \#2}\end{array} &= \begin{array}{c}\text{pipe elevation} \\ \text{at 18{+}00}\end{array} - 1.85 \\
&= 119.76 \text{ ft} - 1.85 \text{ ft} \\
&= 117.91 \text{ ft} \quad [\text{check}]
\end{aligned}$$

Compute the stake elevation.

$$\begin{aligned}
\text{stake elevation} &= \text{pipe elevation} + 10.00 \text{ ft} \\
&= 119.76 \text{ ft} + 10.00 \text{ ft} \\
&= 129.76 \text{ ft}
\end{aligned}$$

Answer (C)

10. Answer (D)

11. Answer (A)

12.
$$\begin{array}{rl}
\text{HI at station 2} = & 102.23 \quad \text{BM} \\
& \underline{+2.12 \quad \text{BS}} \\
& 104.35 \quad \text{HI} \\
& \underline{-11.32 \quad \text{FS @ TP1}} \\
& 93.03 \quad \text{EL @ TP1} \\
& \underline{+1.67 \quad \text{BS}} \\
& 94.70 \quad \text{HI}
\end{array}$$

$$\begin{aligned}
\text{elevation at sta 2{+}00} &= \text{HI} - \text{FS} \\
&= 94.70 \text{ ft} - 4.20 \text{ ft} \\
&= 90.5 \text{ ft}
\end{aligned}$$

Answer (D)

13. Assume that the grade is constant from sta 0+00 to sta 2+00.

$$\begin{aligned}
\text{slope} &= \frac{\text{difference in elevation}}{\text{distance}} \\
&= \frac{91.15 \text{ ft} - 88.00 \text{ ft}}{0{+}75 - 0{+}00} \\
&= \frac{3.15 \text{ ft}}{75 \text{ ft}} \\
&= 0.042
\end{aligned}$$

$$\begin{aligned}
\text{difference in elevation} &= (\text{distance})(\text{slope}) \\
&= (2{+}00 - 0{+}75)(0.042) \\
&= (125 \text{ ft})(0.042) \\
&= 5.25 \text{ ft}
\end{aligned}$$

$$\begin{aligned}
\text{elevation at 2{+}00} &= 91.15 \text{ ft} + 5.25 \text{ ft} \\
&= 96.40 \text{ ft}
\end{aligned}$$

Answer (D)

14. The HI at station 1+75 is 94.70 ft.

$$\begin{aligned}
\text{elevation at sta 1{+}75} &= \text{HI} - \text{FS} \\
&= 94.70 \text{ ft} - 2.1 \text{ ft} \\
&= 92.60 \text{ ft}
\end{aligned}$$

$$\begin{aligned}
\text{grade at sta 1{+}75} &= 91.15 \text{ ft} \\
&\quad + (1{+}75 - 0{+}75)(0.042) \\
&= 91.15 \text{ ft} + 4.2 \\
&= 95.35 \text{ ft}
\end{aligned}$$

$$\begin{aligned}
\text{fill} &= \text{grade} - \text{elevation} \\
&= 95.35 \text{ ft} - 92.60 \text{ ft} \\
&= 2.75 \text{ ft}
\end{aligned}$$

Answer (C)

Photogrammetry 10

1. Introduction

Photogrammetry is taking horizontal and vertical measurements of a site using photography. This chapter will cover *aerial* photogrammetry, which uses aerial photography.

The result of an aerial survey is a topographic map showing contours, spot elevations, and features such as trees, buildings, streets, channels, utilities. To compile such a map, the photogrammetrist uses an instrument known as a *stereoplotter*, wherein the photographs are placed and a stereo picture is produced. From this stereo picture, through a series of manipulations, a topographic map is produced. To achieve the stereo picture of a given area, two overlapping photographs, known as a *model*, of the mapping area must be taken. The photographs are taken from an airplane as it is flying over the site. At least one model is needed for any aerial mapping project, with large projects requiring dozens or even hundreds of models.

2. Ground Control

So that the map scale and leveling of the model can be produced in the stereoplotter, a control network must be established on the ground and must be made visible in each aerial photograph taken along the flight path. In general, a minimum of two horizontally controlled and three vertically controlled points are required for each model; however, five points are suggested to achieve *redundance*. In setting the control points, horizontal control is established in the form of local or state plane coordinates, and vertical control is based on either an assumed or an established vertical datum. The physical control point is a survey monument, which is marked on the ground by a target (or *aerial premark*) large and clear enough to be seen in the aerial photograph. The target usually consists of a white cross painted on the ground (Fig. 10.1), and the size is determined by the scale of the photo: A general rule is to make the length of the target 1/1000 of the photo-scale numerator, in feet, and the width of the target 1/1000 of the photo-scale numerator, in inches. For example, if the photo-scale is 1:7200 or 1 in = 600 ft, the length of the target would be (1/1000)(7200) or 7.2 ft, and the width of the legs of the target would be (1/1000)(7200) or 7.2 in.

Another method of making control points is called *postmarking*. Here, no targets are set prior to photography, but rather the photogrammetrist identifies areas on each photo to be controlled after the photographs are developed and printed. The field crew then physically locates features on the ground that can be seen in the photo. Some features that can be used are traffic striping, angle points in curbs or sidewalks, corners of bridges or retaining walls, prominent rocks, intersections of two trails, and so on. The field crew then circles the feature on the photo and makes a pin prick on the spot being located. Next they tie the point to the horizontal and vertical control being used.

In addition to premarking control points, other features, such as utilities, are targeted for ease of identification on the photo. Figure 10.2 is an example of how certain utilities can be marked. The "+" mark is the center of the feature being marked.

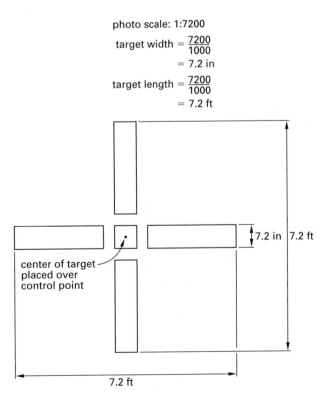

Figure 10.1 *Aerial Target (Premark)*

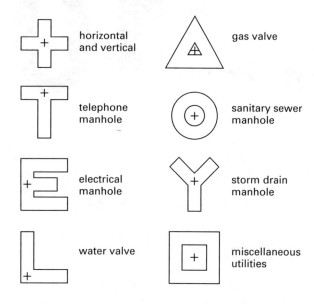

Figure 10.2 *Aerial Control Premark Symbols*

Courtesy of Pfeiler & Associates Engineers, Inc.

3. Photo Scale

A. Scale Expression

The scale of a map or photo may be expressed in several ways. For example, $1'' = 250'$ means 1 in measured on the map is equal to 250 ft measured on the ground.

$1''/250'$ means a ratio of map distance to ground distance where 1 in on the map is equal to 250 ft on the ground.

1:3000 means one unit on the photo equals 3000 units on the ground. If $1'' = 250'$ then $12'' = 3000'$. Since $12'' = 1$ ft, the relative scale, 1:3000, is equivalent to $1'' = 250'$. One inch on the photo equals 3000 in on the ground, 1 yd on the photo equals 3000 yd on the ground, and so on.

1/3000 has the same value as 1:3000.

Therefore, the expressions $1'' = 250'$, $1''/250'$, 1:3000, and 1/3000 state the same scale of a photograph.

Example 10.1

A distance between two features on a photo scaled at $1'' = 250'$ scales to 3.75 in. What is the ground distance between those features?

Solution

Set a ratio as follows.

$$\frac{1}{250 \text{ ft}} = \frac{3.75 \text{ in}}{x}$$

$$x = (3.75 \text{ in})(250 \text{ ft}) = 937.50 \text{ ft}$$

B. Scale Determination

Two factors determine the scale of the photo: focal length of the camera and average flying height above the ground, or altitude. The three common focal length sizes found in aerial cameras are 6 in, 8 in, and 12 in. The most-used size is the 6 in focal length camera. This length provides a 93° field of view, which yields excellent ground coverage with a minimum of control.

Figure 10.3 shows the mathematical relationship of photo-scale, focal length, and altitude.

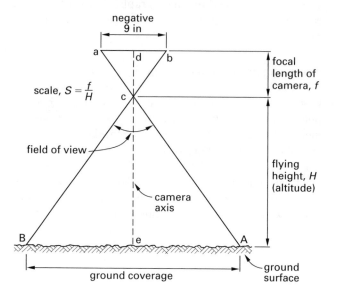

Figure 10.3 *Photo Scale*

ab   the photographic negative, commonly measures 9 in × 9 in

BA   the ground coverage projected onto the negative

c the camera lens

f the focal length of the camera

H the flying height of the plane (altitude)

ce   the camera axis

Triangle abc is similar to triangle ABC since lines ab and BA are parallel and both are perpendicular to de, the axis of the camera. Therefore,

$$\frac{\text{ab}}{\text{BA}} = \frac{\text{dc}}{\text{ce}} \qquad 10.1$$

If $f = $ dc, $H = $ ce, and ab is proportional to BA, then the relative photo-scale can be expressed as a fraction.

$$\frac{1}{x} = \frac{f}{H} \qquad 10.2$$

$$S = \frac{f}{H} \qquad 10.3$$

For most projects, the scale and focal length are fixed quantities. The flying height, H, must be determined.

$$H = \frac{f}{S} \qquad 10.4$$

Example 10.2

Given a focal length of 6 in and a flying height of 3000 ft, find the photo scale.

Solution

Using Eq. 10.3,

$$S = \frac{f}{H} = \frac{6 \text{ in}}{3000 \text{ ft}}$$

If 6 in = 3000 ft, then 12 in (1 foot) = 6000 ft. Therefore, the relative scale can be expressed as

$$1:6000 \text{ or } \frac{1}{6000} \text{ or } 1 \text{ in} = 500 \text{ ft}$$

Example 10.3

Given a desired scale of 1:9600 and a focal length of 6 in, find the flying height.

Solution

Using Eq. 10.3,

$$H = \frac{f}{S} = \frac{0.5}{\dfrac{1}{9600}} = 4800 \text{ ft}$$

Example 10.4

What is the ground coverage attained in Ex. 10.2 in acres?

Solution

Sine the photo measures 9 in × 9 in, the ground coverage is

$$[(9 \text{ in})(500 \text{ ft})][(9 \text{ in})(500 \text{ ft})] = 20{,}250{,}000 \text{ ft}^2$$

$$\frac{20{,}250{,}000 \text{ ft}^2}{43{,}560 \ \dfrac{\text{ft}^2}{\text{ac}}} = 464.88 \text{ ac}$$

4. Overlap

To achieve a stereo picture, an object must show up in at least two successive aerial photographs. This is accomplished by taking the pictures such that they *overlap*. Often on larger projects, the area covered requires more than one flight line, in which case the photographs will overlap in both forward and side directions. Figures 10.4 through 10.6 illustrate the direction of the flight line and the effect of overlap on the ground coverage.

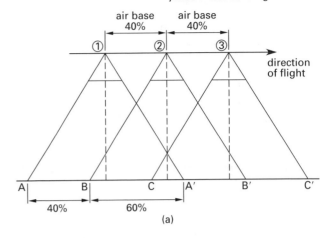

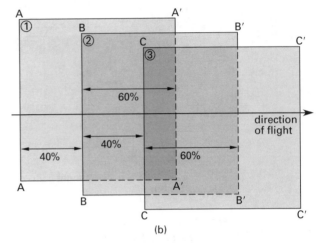

Figure 10.4 Overlap

Since the ground speed of the plane and the height above the mean terrain are known, the *air base* or distance between exposures can be calculated. A device known as an *intervalometer* triggers the camera at a predetermined time interval to yield the desired air base. The percentage of forward overlap is generally 60%, and scale overlap is 30%. When there is more than one flight line, the effective coverage of a 9 in × 9 in photograph is

$$(9 \text{ in})(0.40) = 3.6 \text{ in forward}$$
$$(9 \text{ in})(0.70) = 6.3 \text{ in side}$$

This coverage is called the *neat model*. In Fig. 10.4(b), the neat model is illustrated as the portion of photograph 2 between the left edge of photograph 2 and the left edge of photograph 3. This portion of the photograph would measure 3.6 in along the flight line. If there is more than one flight line, the side coverage for photograph 2, as seen in Fig. 10.5(b), is between the right edge of photograph 1 and the left edge of photograph 3.

side overlap—generally: 30%
yields : 70% coverage

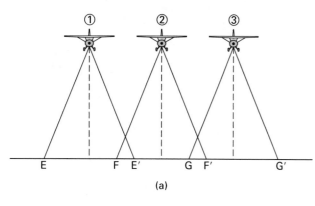

(a)

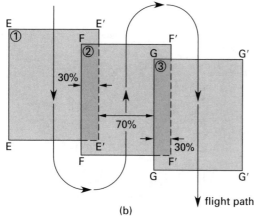

(b)

Figure 10.5 *Side Overlap*
(more than one flight line)

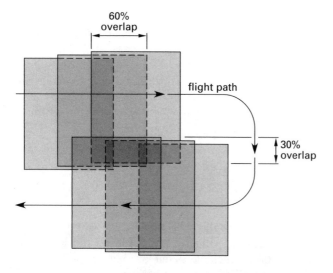

Figure 10.6 *Side and Forward Overlap*
for a Single Flight Path

At a photo scale of 1 in = 1200 ft, the effective coverage of the neat model would be computed as follows.

$$[(9 \text{ in})(1200 \text{ ft})](0.40) = 4320 \text{ ft} \quad [\text{forward}]$$

$$[(9 \text{ in})(1200 \text{ ft})](0.70) = 7560 \text{ ft} \quad [\text{side}]$$

$$\frac{(4320 \text{ ft})(7560 \text{ ft})}{43,560} = 749.8 \text{ ac}$$

5. Mapping Criteria

Prior to flying a project, the designer must provide the photogrammetrist with the desired map scale and contour interval. Most engineering design projects, such as highways, storm drains, or street improvements, are mapped at 1 in = 50 ft. Drainage studies, flood control maps, and tentative maps are mapped at a scale of 1 in = 100 ft or 1 in = 200 ft. Planning projects such as preliminary route design are flown at a scale of 1 in = 1000 ft to 1 in = 2000 ft. Most designs for improvements require that the contour interval be 1 ft or 2 ft. Other studies may utilize maps having a contour interval of 4 ft, 5 ft, or 10 ft. For large-scale drainage studies, 20 ft contours are acceptable.

Every stereoplotter has an inherent factor called the *C-factor*, which governs the contour interval that can be plotted. The C-factor can be defined as "the allowable flying height for a 1 ft contour interval." The formula for computing the contour interval (CI) is

$$\text{CI} = \frac{H}{\text{C-factor}} = \text{contour interval} \qquad 10.5$$

$$H = \text{altitude}$$

The C-factor is given for the individual stereoplotter.

Conversely, the flying height can be computed as follows.

$$H = (\text{CI})(\text{C-factor}) \qquad 10.6$$

The standard C-factor for most plotters in use today is 1500.

Example 10.5

Given a plotter C-factor of 1200, a desired CI of 2 ft, and a focal length of 6 in, find the altitude (H) and the photo scale.

Solution

Find the altitude (H) by using Eq. 10.6.

$$H = (\text{CI})(\text{C-factor}) = (2 \text{ ft})(1200) = 2400 \text{ ft}$$

Find the photo scale by using Eq. 10.3.

$$S = \frac{f}{H} = \frac{6 \text{ in}}{2400 \text{ ft}} = \frac{1}{400} \text{ or } 1 \text{ in} = 400 \text{ ft}$$

Another factor inherent in the plotter is called a *diameter factor* (also called *D-factor*), which is used to determine the map scale for each photo scale. Most plotters have a D-factor of 5, where the map scale is five times the photo scale. In Ex. 10.5, the map scale would be five times the photo scale, or 1 in = 80 ft. Equation 10.7 is used for determining map scale.

$$\text{map scale} = \frac{\text{photo scale}}{\text{D-factor}} = \frac{400 \, \frac{\text{ft}}{\text{in}}}{5}$$

$$= 80 \text{ ft/in or } 1 \text{ in} = 80 \text{ ft} \qquad 10.7$$

6. Planning the Flight

Although the photogrammetrist is responsible for flight planning, the project manager should be familiar with the process in order to aid in the preliminary planning phase of the project. The engineer will most likely be asked to determine the cost of the project so that a budget can be prepared. To determine the cost of a project, the number of neat models required must be computed.

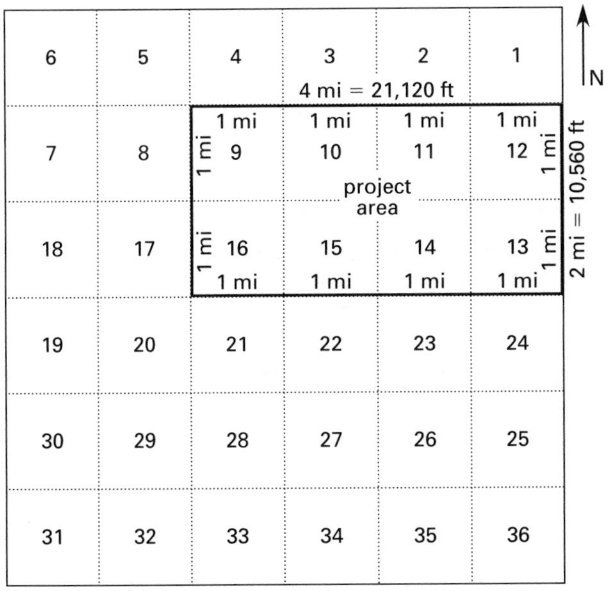

Figure 10.7 *Flight Planning*

Example 10.6

Set up the flight information needed to map a project covering eight sections of public land in T 15 N, R 10 W, XYZ meridian. The sections are 9, 10, 11, 12, 13, 14, 15, and 16. Assume that the dimensions of each section are 1 mi × 1 mi or 5280 ft × 5280 ft. The client desires a map showing 4 ft contours. The photogrammetrist will be compiling the map using a plotter with a C-factor of 1500 and a camera with a 6 in focal length.

Solution

Determine the flying height needed to yield a 4 ft contour interval using Eq. 10.6.

$$H = (\text{CI})(\text{C-factor}) = (4 \text{ ft})(1500) = 6000 \text{ ft}$$

Determine the photo scale using Eq. 10.3.

$$S = \frac{f}{H} = \frac{6 \text{ in}}{6000 \text{ ft}} = \frac{1 \text{ in}}{1000 \text{ ft}} \text{ or } 1 \text{ in} = 1000 \text{ ft}$$

Determine the forward and side coverage.

Assume that the forward overlap will be 60% and the side overlap will be 30%.

$$\text{forward coverage} = (9 \text{ in})(40\%) = 3.6 \text{ in per photo}$$

At 1 in = 1000 ft, the forward coverage equals (3.6 in)(1000 ft) = 3600 ft.

$$\text{side coverage} = (9 \text{ in})(70\%) = 6.3 \text{ in per photo}$$

At 1 in = 1000 ft, the side coverage is equal to (6.3 in)(1000 ft) = 6300 ft.

Determine the number of flight lines.

The width of the project is (5280 ft)(2) = 10,560 ft. The side coverage is 6300 ft per model. Thus,

$$\text{no. of flight lines} = \frac{\text{width}}{\text{coverage}} = \frac{10,560 \text{ ft}}{6300 \text{ ft}}$$

$$= 1.7 \quad (\text{say } 2)$$

Determine the number of models required.

The length of the project is (5280 ft)(4) = 21,120 ft. The forward coverage is 3600 ft per model. Therefore,

$$\text{no. of models} = \frac{\text{length}}{\text{coverage}} = \frac{21,120 \text{ ft}}{3600 \, \frac{\text{ft}}{\text{model}}}$$

$$= 5.9 \quad [6 \text{ photos per flight line}]$$

There are two flight lines; therefore the number of models is 12.

Determine the number of photos that will be required to provide stereoscopic coverage.

To achieve stereo, the number of photos required will be the number of models, n, plus 1 per flight line.

$$\text{no. of photos} = (n+1)(2) = (6+1)(2)$$

$$= 14 \text{ photos required}$$

7. Horizontal and Vertical Control

The photogrammetrist will plan a *control scheme* showing the number and location of the targets required to control the plotting of the map. Figure 10.8 shows the ideal control location for a 1-model and a 3-model project.

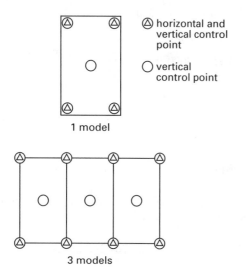

Figure 10.8 *Horizontal and Vertical Control*

The 1-model project requires that five control points be furnished. The 3-model project also requires that five control points per model be set; however, two of the points are common to adjoining models. The actual number of points for 3 models is 11, not 15. The control scheme necessary for the project outlined in Ex. 10.6 is shown in Fig. 10.9.

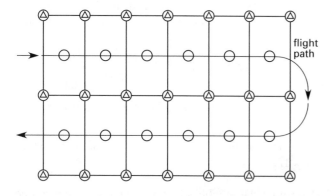

Figure 10.9 *Flight Path Using Horizontal and Vertical Control*

The control points between the two tiers of models are common to both; therefore, the number of targets needed to control the 12 models would be 33.

8. Flight Plan

Prior to actually flying a project and taking aerial photographs, the photogrammetrist prepares a flight plan and gives it to the flight crew and to the person establishing the control on the ground. The flight plan shows the area to be mapped, the direction of flight and flight lines, the desired target locations, the height above mean terrain (AMT), and the height above sea level (ASL). Figure 10.10 shows the relationship of AMT and ASL.

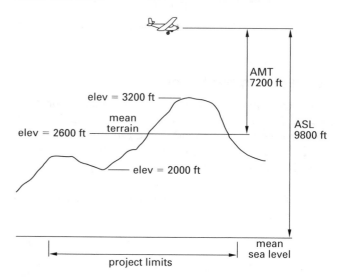

Figure 10.10 *Typical Photogrammetric Flight*

Example 10.7

Given a focal length of 6 in and a photo scale of 1 in = 1200 ft or 1:14,400, find AMT and ASL.

Solution

The mean terrain is the average elevation over the project and is computed by averaging the high and low elevations. H (the flying height) is derived from Eq. 10.4.

$$H = \frac{f}{S} = \frac{6 \text{ in}}{\dfrac{1}{14,400} \dfrac{\text{in}}{\text{ft}}}$$

$$= 7200 \text{ ft} \quad [\text{AMT}]$$

The ASL is computed by adding the flying height to the elevation of the mean terrain. In Fig. 10.10, the ASL is computed as follows.

$$\text{ASL} = \text{AMT} + \text{elevation of mean terrain}$$
$$= 7200 \text{ ft} + 2600 \text{ ft}$$
$$= 9800 \text{ ft}$$

This is the elevation at which the plane must fly to obtain the desired scale. The plane is equipped with instrumentation that will allow the ASL to be maintained throughout the flight.

9. Map Accuracy

The following standards of accuracy must be met.

A. Horizontal Accuracy

- For maps with publication scales larger than 1:20,000, not more than 10% of the points tested shall be in error by more than 1/30 of an inch.

- For maps with publication scales of 1:20,000 or smaller, not more than 10% of the points tested shall be in error by more than 1/50 of an inch.

A map scale is larger than another map scale if the denominator of the relative scale is smaller—that is, a map whose scale is 1/12,000 has a larger map scale than one whose scale is 1/24,000.

B. Vertical Accuracy

- *Contours*—at any scale, not more than 10% of the elevations tested shall be in error of more than one-half the contour interval, and none can exceed the interval.

- *Spot elevations*—90% of all spot elevations shall be accurate to within one-fourth the contour interval, while the remaining 10% shall not be in error of more than one-half the contour interval.

Practice Problems

Problems 1 through 12 are based on the following information.

Project data: Sections 9 and 16 and the westerly 4000 ft of Sections 10 and 15, T 6 S, R 24 W, S.B.M.

$$
\begin{array}{rl}
\text{focal length:} & \text{6 in} \\
\text{size of negative:} & \text{9 in} \times \text{9 in} \\
\text{contour interval (CI):} & \text{5 ft} \\
\text{forward overlap:} & \text{60\%} \\
\text{side overlap:} & \text{22\%} \\
\text{C-factor:} & \text{1800 ft} \\
\text{D-factor:} & \text{5 to 1} \\
\text{average terrain:} & \text{2500 ft above sea level}
\end{array}
$$

1. What is the flying height (H)?
   - (A) 360 ft
   - (B) 1800 ft
   - (C) 9000 ft
   - (D) 11,500 ft

2. What is the required flying height above sea level (ASL)?
   - (A) 1800 ft
   - (B) 9000 ft
   - (C) 10,800 ft
   - (D) 11,500 ft

3. What is the photo scale?
   - (A) 1 in = 1500 ft
   - (B) 1 in = 1800 ft
   - (C) 1 in = 3000 ft
   - (D) 1 in = 18,000 ft

4. What is the map scale?
   - (A) 1 in = 300 ft
   - (B) 1 in = 360 ft
   - (C) 1 in = 600 ft
   - (D) 1 in = 3600 ft

5. How many flight lines are there?
   - (A) 1
   - (B) 2
   - (C) 3
   - (D) 4

6. What is the minimum number of models required?
   - (A) 1
   - (B) 2
   - (C) 3
   - (D) 4

7. What is the minimum number of photographs required?
   - (A) 1
   - (B) 2
   - (C) 3
   - (D) 4

8. What should be the minimum length of the aerial target?
   - (A) 15 ft
   - (B) 16 ft
   - (C) 17 ft
   - (D) 18 ft

9. What should be the minimum width of the aerial target?

  (A) 15 in
  (B) 16 in
  (C) 17 in
  (D) 18 in

10. Ninety percent of the contours plotted on the map must be accurate to within which of the following?

  (A) 1.0 ft
  (B) 2.0 ft
  (C) 2.5 ft
  (D) 5.0 ft

11. Ninety percent of the spot elevations plotted on the map must be accurate to within which of the following?

  (A) 1.0 ft
  (B) 1.25 ft
  (C) 2.5 ft
  (D) 5.0 ft

12. The positional accuracy of 90% of the planimetric features plotted on the map must be accurate to within which of the following?

  (A) 3.0 ft
  (B) 6.0 ft
  (C) 10.0 ft
  (D) 30.0 ft

Solutions

1. To find the flying height, use Eq. 10.6. The C-factor and contour interval are given.

$$H = \text{(C-factor)(CI)}$$
$$= (1800 \text{ ft})(5)$$
$$= 9000 \text{ ft}$$

Answer (C)

2. The required flying height above sea level (ASL) is the sum of the flying height and the elevation of the average terrain.

$$\text{ASL} = 9000 \text{ ft} + 2500 \text{ ft} = 11{,}500 \text{ ft}$$

Answer (D)

3. The photo scale is a function of the flying height and the focal length of the camera.

$$S = \frac{f}{H} = \frac{6 \text{ in}}{9000 \text{ ft}} = \frac{0.5 \text{ ft}}{9000 \text{ ft}}$$
$$= \frac{1 \text{ ft}}{18{,}000 \text{ ft}} \text{ or } 1 \text{ in}$$
$$= \frac{18{,}000 \text{ ft}}{12 \dfrac{\text{in}}{\text{ft}}}$$
$$= 1500 \text{ ft}$$

Answer (A)

4. The map scale is a function of the photo scale and the D-factor.

$$\text{map scale} = \frac{\text{photo scale}}{\text{D-factor}} = \frac{1500 \dfrac{\text{ft}}{\text{in}}}{5}$$
$$= 300 \text{ ft/in or } 1 \text{ in} = 300 \text{ ft}$$

Answer (A)

5. The east-west project dimensions are 9280 ft, since one section of land is 5280 ft on a side and the adjacent piece is 4000 ft. The north-south project dimensions are 10,560 ft since the project covers two sections, north and south. The flight will usually be made along the largest dimension, in this case north to south or south to north. With a 22% overlap, the net coverage will be 78% of the photograph, yielding 7.02 in (i.e., (9 in)(0.78) = 7.02 in). At a scale of 1 in = 1500, the net ground coverage will be

$$\text{coverage} = (7.02 \text{ in}) \left(1500 \dfrac{\text{ft}}{\text{in}} \right)$$
$$= 10{,}530 \text{ ft}$$

The entire project is 9280 ft across; therefore, the amount covered in one pass or flight will be sufficient.

$$\frac{\text{project width}}{\text{coverage}} = \frac{9280 \text{ ft}}{10{,}530 \text{ ft}}$$
$$= 0.88 \quad (\text{say } 1)$$

Answer (A)

6. The number of models is a function of the net amount covered by one photograph and the total length of the project.

At 60% forward overlap, the net coverage will be 40% of the photograph, yielding 3.6 in (i.e., (9 in)(0.4) = 3.6 in). At a photo scale of 1 in = 1500 ft, the net ground coverage or model length will be

$$\left(1500 \ \frac{\text{ft}}{\text{in}}\right)(3.6 \text{ in}) = 5400 \text{ ft}$$

The project is 10,560 ft long; therefore by dividing the project length by the model length, the number of models can be found.

$$\frac{10{,}560 \text{ ft}}{5400 \text{ ft}} = 1.95 \quad (\text{say } 2)$$

Answer (B)

7. To achieve the stereoscopic effect needed to plot a map, there must be two photos per model; however, since the photos overlap, each photo is used in the successive model, thus the number of photographs needed for a mapping project is the number of models plus one. In this problem, there are two models, therefore three photographs are required.

Answer (C)

8. The length of the aerial target, in feet, is computed by multiplying the denominator of the relative photo scale by 0.001. In this problem, the relative photoscale is 1/18,000, so the length of the target equals (18,000)(0.001) = 18 ft.

Answer (D)

9. The width of the aerial target, in inches, is computed by multiplying the denominator of the relative photo scale by 0.001 ft. The width is (18,000)(0.001) = 18 in.

Answer (D)

10. Ninety percent of the contours plotted on a map must be accurate to within one-half of the contour interval (CI) of the map. The CI in the problem is 5 ft; therefore the answer is 2.5 ft.

Answer (C)

11. Ninety percent of the spot elevations plotted on a map must be accurate to within one-fourth of the contour interval of the map. Therefore the answer is 1.25 ft.

Answer (B)

12. The positional accuracy of a planimetric feature plotted on this map must be within 1/50 of an inch. At a map scale of 1 in = 300 ft, this will be 6.0 ft.

Answer (B)

California Coordinate System

11

Contributed by Jeremy Evans, PLS

1. Introduction

The *California Coordinate System* (CCS) is a system of plane rectangular coordinates that has been established by California statutes for defining and stating the positions of points on the surface of the earth within the state. Currently there are two CCSs used in California: the *California Coordinate System of 1927* (CCS27) and the *California Coordinate System of 1983* (CCS83). Calculations involving *state plane coordinate* (SPC) systems include the following.

- transformation of coordinates between the reference ellipsoid (latitude and longitude) and the grid (northings and eastings)

- the adjustment of survey date (lengths and azimuths) in converting between ground, ellipsoid, and grid

- transformation from one SPC zone to another

- coordinate geometry with adjusted observations on the SPC grid

2. Origin

The CCS was established by the *United States Coast and Geodetic Survey* (now known as the *National Geodetic Survey*, NGS). Legislation enacting the CCS was passed in 1947 (CCS27) and 1986 (CCS83). CCS27 is being phased out, and all state plane work after January 1, 1995 refers to CCS83.

3. Benefits

The CCS benefits all surveyors and engineers by providing the following.

- a method for determining the plane coordinates (northing and easting) of a point from its geodetic coordinates (latitude and longitude)

- the use of plane survey techniques and calculations over large areas that do not introduce significant errors

- a single reference system for all surveys in an area (With the advent of Geographical Information Systems (GIS) where data from a large area must be collected and referenced to a common datum, CCS is a must.)

- a reference system that allows for easy retracing of survey work done previously on the CCS; a uniform computational base

4. CCS27 versus CCS83

The basic differences between the two systems are:

- CCS27 is referenced to Clarke spheroid of 1866, while CCS83 is referenced to the Geodetic Reference System of 1980.

- The unit of measure for CCS27 is the U.S. survey foot, while the unit of measure for CCS83 is the meter.

- The location of the point of origin for CCS27 and CCS83 is different; therefore, the place coordinates of a station have significantly different values.

- Because the reference datum differs between the two systems, the CCS27 latitude and longitude of a station are different from the CCS83 latitude and longitude. This difference is not constant throughout the state; therefore, a simple transformation of coordinates between the two systems in not possible.

- CCS27 divides the state into seven CCS zones, while CCS83 divides it into six. (See Figs. 11.1 and 11.2.)

5. Shape of the Earth

The earth is normally thought of as a sphere, but actually, it is flattened slightly at the poles, assuming the shape of a mathematical figure called an *oblate spheroid* (see Fig. 11.3).

Because of the earth's surface variations, computations cannot be made based on its true surface because of the difficulty in deriving equations. Geodesists (those who study the shape of the earth) have developed numerous mathematical surfaces, called *spheroids* and *ellipsoids*, which closely resemble the surface of the earth at sea level (Fig. 11.4). These surfaces, or datum, allow the surveyor to determine the location of points and perform calculations involving these points. Calculations done on these surfaces are *geodetic calculations*.

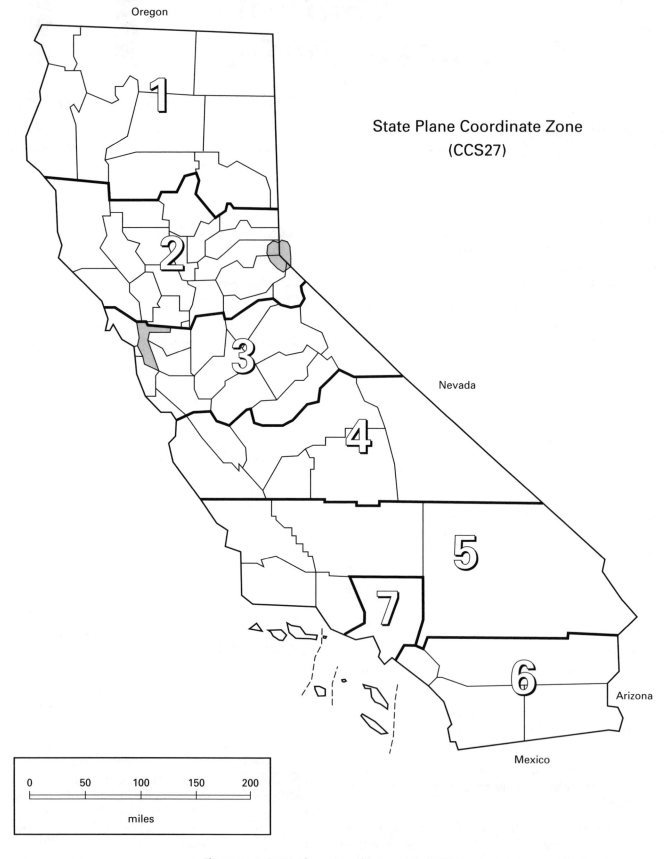

Figure 11.1 *State Plane Coordinate Zone (CCS27)*

Reprinted with permission from Roy Minnick, NAD 83, © 1989 Landmark Enterprises.

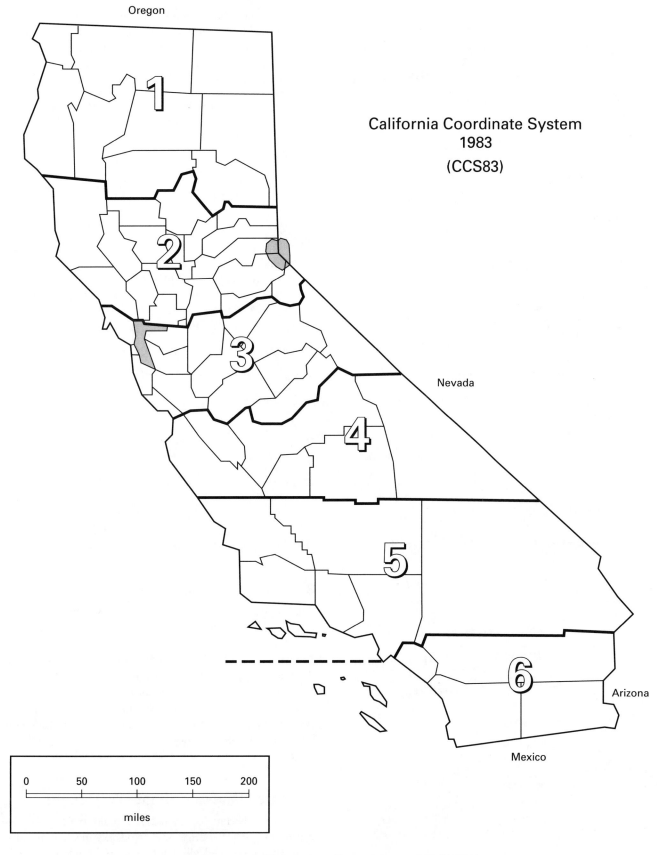

Oregon

California Coordinate System
1983

(CCS83)

Nevada

Arizona

Mexico

0	50	100	150	200

miles

Figure 11.2 *California Coordinate System 1983 (CCS83)*

Reprinted with permission from Roy Minnick, NAD 83, © 1989 Landmark Enterprises.

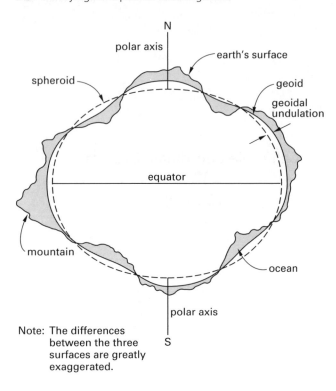

Note: The differences between the three surfaces are greatly exaggerated.

Figure 11.3 Shape of the Earth

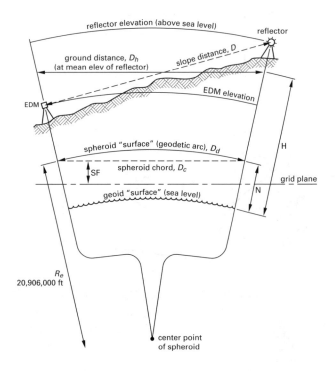

Figure 11.4 Mathematical Surfaces

6. Clarke Spheroid of 1866

The reference datum for CCS27 is the Clarke spheroid of 1866. This datum is considered a regional datum since it is a best fit to North America and cannot be used in the rest of the world. The dimensions of the Clarke spheroid are as follows.

- equatorial radius: 20,925,831 (U.S. survey feet)

- polar radius: 20,854,891 ft

- flattening: 1/295.0

$$\text{flattening} = \frac{\text{equatorial radius} - \text{polar radius}}{\text{equatorial radius}}$$

The reference ellipsoid for CCS83 is the *Geodetic Reference System of 1980* (GRS80). This datum is a global datum since the center of the ellipsoid is at the center of the earth's mass. Calculations using GRS80 are equally accurate anywhere on the earth. The dimensions of GRS80 are as follows.

- equatorial radius: 20,925,604 ft (U.S. survey feet)

- polar radius: 20,855,445 ft

- flattening: 1/298.257...

7. Geoid

The geoid is the irregular surface of variable radius that would exist if the earth was covered with water. It is considered *sea level*. The varying radius is caused by local variations in the direction of gravity giving the geoid a lumpy potato appearance. Survey data referenced to the geoid is called *astronomic*.

8. Position on the Earth

The position of a point on the earth's surface can be referenced as place names, plane coordinates, and so on. The spherical coordinates of a point are expressed as latitude and longitude. The *latitude* of a point is the distance, measured in degrees, north or south from the equator to the point. The *longitude* of a point is the distance, measured in degrees, east or west from the prime (Greenwich) meridian. Both latitude and longitude can be expressed as *astronomic* (relating to the geoid) or *geodetic* (relating to the reference spheroid/ellipsoid).

9. State Plane Coordinate Projections

To convert geodetic coordinates to *plane rectangular coordinates* (SPC), points are projected mathematically from the referenced ellipsoid to some imaginary developable surface. This surface can be unrolled and layed flat without distortion of shape or size. A rectangular

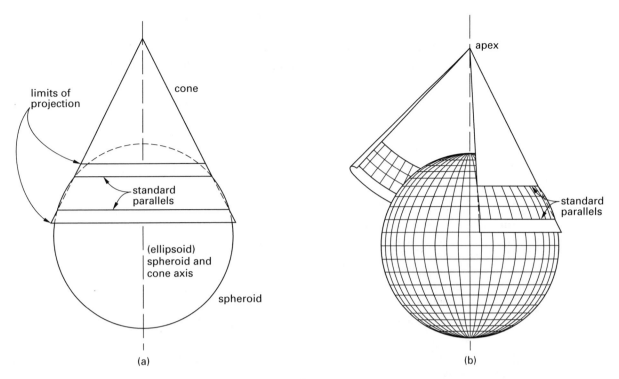

Figure 11.5 *Lambert Projection*

grid is then superimposed on the developed plane surface, and the positions of the points can be determined on the grid plane. A grid plane developed using this mathematical process is called a *map projection*.

10. Lambert Conformal Conic Projection

The *Lambert conformal conic projection* is a projection onto the surface of an imaginary cone. *Conformal* means that angular relationships are retained around all points when projected from the ellipsoid to the cone. The *Lambert projection* is relatively distortion-free in an east-west direction, and distortion is minimized in a north-south direction by limiting each zone to about 158 mi wide.

In the Lambert projection, the cone intersects the ellipsoid along two parallels of latitude at one-sixth of the zone width from the north and south limits. These two parallels are called *standard parallels*. As shown in Fig. 11.5, on the Lambert projection all parallels of latitude are arcs of concentric circles having the centers at the apex of the cone. The meridians are straight lines that converge at the apex. The central meridian is near the middle of the zone, and it determines grid north. This is the only meridian where true (astronomic) north and grid north are the same.

As seen in Fig. 11.6, directions of true and grid north do not coincide except at the central meridian. The difference between the two is called the *convergence angle*.

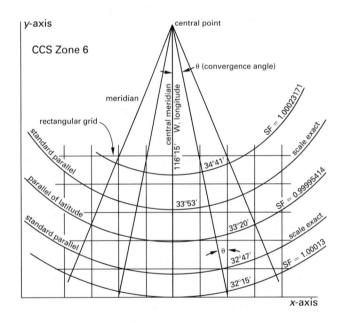

Figure 11.6 *Parallels, Latitudes, and Meridians*

11. Calculations on the Lambert Projection

The Lambert projection mathematically converts geodetic coordinates to grid coordinates. This transformation can be done using several methods, including projection tables.

Tables 11.1 and 11.2 are a portion of the projection tables for California.

Table 11.1 California Zone 6, CA06, Zone #046

California Zone 6, CA06, Zone #0406
North American Datum 1983 (NAD83)—California Coordinate System 1983 (CCS83)

meters		U.S. survey feet	
B_s	= 32°47′ N	B_s	= 32°47′ N
B_n	= 33°53′ N	B_n	= 33°53′ N
B_b	= 32°10′ N	B_b	= 32°10′ N
L_o	= 116°15′ W = CM	L_o	= 116°15′ W
N_b	= 500000.0000 m	N_b	= 1640416.667′
E_o	= 2000000.0000 m	E_o	= 6561666.667′
B_o	= 33.3339229447° N	B_o	= 33.3339229447° N
$\sin B_o$	= 0.549517575763 = l	$\sin B_o$	= 0.549517575763
R_b	= 9836091.7896 m	R_b	= 32270577.813′
R_o	= 9706640.0762 m	R_o	= 31845868.317′
N_o	= 629451.7134 m	N_o	= 2065126.163′
K	= 13602026.7133 m	K	= 44625982.642′
k_o	= 0.999954142490	k_o	= 0.999954142490
M_o	= 6354407.2007 m	M_o	= 20847750.958p
r_o	= 6369336 m	r_o	= 20896729.860′
L_1	= 110905.3274 m	L_1	= 363861.8950
L_2	= 8.94188 m	L_2	= 29.3368
L_3	= 5.65087 m	L_3	= 18.5396
L_4	= 0.016171 m	L_4	= 0.053054
G_1	= 9.016699372E−06	G_1	= 2.748295465E−06
G_2	= −6.55499E−15	G_2	= −6.08981E−16
G_3	= −3.73318E−20	G_3	= −1.05713E−21
G_4	= −8.2753E−28	G_4	= −7.1424E−30
F_1	= 0.999954142490	F_1	= 0.999954142490
F_2	= 1.23251E−14	F_2	= 1.14504E−15
F_3	= 4.15E−22	F_3	= 1.18E−23

The customary limits of the zone are from 32°10′ N to 34°30′ N.

Reprinted with permission from Vincent J. Sincek, *CCS83—A Practical Guide*, © 1992.

These tables are available from the NGS and the California Land Surveyors Association. The tables list the data necessary to do the coordinate transformation from geodetic to grid and vice versa. The data is tabulated for every minute of latitude within the zone. For points not exactly on an even minute of latitude, the data can be interpolated for the exact position.

To generate grid coordinates from geodetic coordinates (and vice versa), solve the conversion triangle shown in Fig. 11.7.

Point P is shown with both its geodetic coordinates (latitude and longitude) and its grid coordinates (northings and eastings).

R_b is the linear distance (measured in meters for CCS83) between the apex of the cone and the southernmost latitude of the zone. This value is a constant for each zone and is given in the projection tables.

CM is the longitude of the central meridian of the zone and is given in the projection tables.

R is the linear distance between the apex of the cone and point P and is determined from the latitude of point P. This distance is determined by interpolation of the values of R given in the projection tables.

The *convergence angle* (CA) is the angle at the apex of the cone between the meridian of the central meridian and the meridian of point P, determined from the longitude of these two meridians.

N_b is the northing of the origin of the grid and has a value of 500 000 m for each zone. The origin of the grid is the intersection of the central meridian and the southernmost latitude of the zone.

E_o is the easting of the origin of the grid and has a value if 2 000 000 m for each zone.

Table 11.2 *Zone 6 California Mapping Radius*

Zone 6 California Mapping Radius
North American Datum 1983

$$K = 13602026.7133 \qquad L = 0.549517575763 \qquad e = 0.0818191910428$$

33°

min	R (meter)	tab diff (per sec)	R (US survey ft)	tab diff (per sec)	scale factor
		30.80699		101.07261	
20	9,706,705.4672		31,846,082.85		0.99995414
		30.80708		101.07288	
21	9,704,857.0427		31,840,018.48		0.99995418
		30.80716		101.07315	
22	9,703,008.6132		31,833,954.09		0.99995431
		30.80725		101.07345	
23	9,701,160.1783		31,827,889.68		0.99995451
		30.80734		101.07374	
24	9,699,311.7380		31,821,825.26		0.99995480
		30.80743		101.07405	
25	9,697,463.2921		31,815,760.82		0.99995518
		30.80753		101.07436	
26	9,695,614.8405		31,809,696.36		0.99995564
		30.80763		101.07468	
27	9,693,766.3830		31,803,631.87		0.99995619
		30.80773		101.07502	
28	9,691,917.9193		31,797,567.37		0.99995682
		30.80783		101.07536	
29	9,690,069.4495		31,791,502.85		0.99995753
		30.80794		101.07571	
30	9,688,220.9733		31,785,438.31		0.99995833
		30.80804		101.07606	
31	9,686,372.4906		31,779,373.75		0.99995921
		30.80816		101.07643	
32	9,684,524.0012		31,773,309.16		0.99996018
		30.80827		101.07680	
33	9,682,675.5050		31,767,244.55		0.99996123
		30.80839		101.07719	
34	9,680,827.0017		31,761,179.92		0.99996236
		30.80851		101.07758	
35	9,678,978.4913		31,755,115.27		0.99996358
		30.80863		101.07798	
36	9,677,129.9735		31,749,050.59		0.99996489
		30.80875		101.07839	
37	9,675,281.4482		31,742,985.88		0.99996627
		30.80888		101.07880	
38	9,673,432.9115		31,736,921.16		0.99996775
		30.80901		101.07923	
39	9,671,584.3748		31,730,856.40		0.99996931

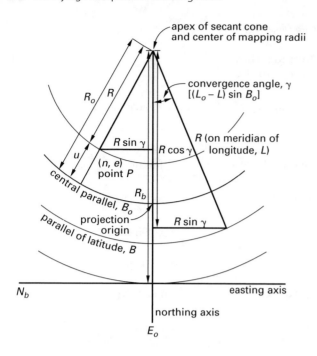

Figure 11.7 *Conversion Triangle*

Reprinted with permission from Vincent J. Sincek, *CCS83—A Practical Guide*, © 1992.

L is a conversion factor used in generating the CA and is a constant for each zone. It is listed in the projection tables.

The northing of point P can be determined from the following formula.

$$\text{northing P} = R_b - R\cos\text{CA} + N_b$$

The easting can be determined from the following formula.

$$\text{easting P} = R\sin\text{CA} + E_o$$

Example 11.1

Given the following geodetic coordinates of point P,

$$\text{latitude} = 33°27'51.19456'' \text{ N}$$
$$\text{longitude} = 117°42'33.35756'' \text{ W}$$

Determine the CCS83 Zone 6 coordinates of point P.

Solution

From Tables 11.1 and 11.2, the following constants are given.

$$R_b = 9\,836\,091.7896 \text{ m}$$
$$N_b = 500\,000 \text{ m}$$
$$E_o = 2\,000\,000 \text{ m}$$
$$\text{CM} = 116°15'$$
$$L = 0.549517575763$$

From Table 11.2, the value of R for latitude 33°27′ is 9 693 766.3830 m. The tabular difference between latitude 33°27′ and latitude 33°28′ is 30.80773 m. The tabular difference is the linear distance for 1 sec of latitude and is listed between every minute of latitude in the projection tables. To determine the linear distance between point P and latitude 33°27′, multiply the seconds of the latitude of point P by the tabular difference.

$$\text{distance} = (51.19456)(30.80773 \text{ m})$$
$$= 1577.1882 \text{ m}$$

Subtract this value from the R value for latitude 33°27′ to determine the R value for point P.

$$R = 9\,693\,766.3830 \text{ m} - 1577.1882 \text{ m}$$
$$= 9\,692\,189.1948 \text{ m}$$

To determine the CA for point P, determine the difference in longitude between the CM and point P.

$$\text{difference in longitude} = 116°15' - 117°42'33.35756''$$
$$= -1°27'33.35756''$$
$$= -1.4592659889°$$

[the algebraic sign is significant]

This difference in longitude is multiplied by the constant L to determine CA.

$$\text{CA} = (-1.4592659889°)(0.549517575763)$$
$$= -0.8018923086°$$
$$= -0°48'06.8123''$$

The values for the other side of the conversion triangle are

$$R\cos\text{CA} = (9\,692\,189.1948 \text{ m})\cos(-0°48'06.8123'')$$
$$= 9\,691\,239.9644 \text{ m}$$

$$R\sin\text{CA} = (9\,692\,189.1948 \text{ m})\sin(-0°48'06.8123'')$$
$$= -135\,644.1662 \text{ m}$$

Calculate the northing and easting of point P.

$$\text{northing P} = R_b - R\cos\text{CA} + N_b$$
$$= 9\,836\,091.7896 \text{ m} - 9\,691\,239.9644 \text{ m}$$
$$+ 500\,000 \text{ m}$$
$$= 644\,851.8252 \text{ m}$$
$$= 644\,851.825 \text{ m}$$

$$\begin{bmatrix}\text{final answers should be rounded} \\ \text{to three decimal places}\end{bmatrix}$$

The answer in U.S. survey feet is

$$\text{northing P} = (644{,}851.8252)\left(\frac{39.37 \text{ m}}{12 \text{ in}}\right)$$
$$= 2{,}115{,}651.36 \text{ ft}$$

$$\text{easting P} = E_o + R\sin\text{CA}$$
$$= 2\,000\,000 \text{ m} + (-135\,644.1662 \text{ m})$$
$$= 1\,864\,355.8338 \text{ m}$$
$$= 1\,864\,355.835 \text{ m}$$
$$= 6{,}116{,}640.76 \text{ ft}$$

There is no value for N_b, and the coordinates at the origin are $x = 2{,}000{,}000$ ft and $y = 0$ ft.

Example 11.2

Transform the grid coordinates for point P back to geodetic.

Solution

Determine the two shortest sides of the conversion triangle.

$$R\cos\text{CA} = R_b - \text{northing P} + N_b$$
$$= 9\,836\,901.7896 \text{ m} - 644\,851.8252 \text{ m}$$
$$\qquad + 500\,000 \text{ m}$$
$$= 9\,691\,239.9644 \text{ m}$$

$$R\sin\text{CA} = \text{easting P} - E_o$$
$$= 1\,864\,355.8338 \text{ m} - 2\,000\,000 \text{ m}$$
$$= -135\,644.1662 \text{ m}$$

Calculate R and CA for point P.

$$R^2 = (R\sin\text{CA})^2 + (R\cos\text{CA})^2$$
$$= (135\,644.1662 \text{ m})^2 + (9\,691\,239.9644 \text{ m})^2$$

$$R = 9\,692\,189.1948 \text{ m}$$

$$\tan\text{CA} = \frac{R\sin\text{CA}}{R\cos\text{CA}}$$
$$= \frac{-135\,644.1662 \text{ m}}{9\,691\,239.9664 \text{ m}}$$
$$= -0.0139965749 \text{ m}$$

$$\text{CA} = -0.8018923086°$$
$$= -0°48'06.8123''$$

From the projection tables, the value of R for point P is between $33°27'$ and $33°28'$ latitude. The tabular difference between these two values is 30.80773 m. To determine the latitude of point P, determine the linear distance between latitude $33°27'$ and point P. Then use the tabular difference to convert this distance to seconds of latitude.

$$\text{linear distance} = 9\,693\,766.3830 \text{ m}$$
$$\qquad - 6\,692\,189.1948 \text{ m}$$
$$= 1577.1882 \text{ m}$$

$$\text{angular distance} = \frac{1577.1882 \text{ m}}{30.80773 \text{ m}}$$
$$= 51.19456'$$

$$\text{latitude} = 33°27' + 51.19456'$$
$$= 33°27'51.19456''$$

To determine the longitude of point P, the CA is divided by the value of L to determine the difference between the meridian of the central meridian and the meridian of point P.

$$\text{difference} = \frac{\text{CA}}{L}$$
$$= \frac{0.8018923086°}{0.549517575763}$$
$$= -1°27'33.35756''$$

The longitude of point P is

$$\text{longitude} = 116°15' - (-1°27'33.35756'')$$
$$= 117°42'33.35756''$$

12. Traversing Using State Plane Coordinates

The basics of traversing were discussed in Chap. 6. Traversing using SPC is the same except that ground distances must be converted to grid distances and azimuths must be converted to grid azimuths. After these two conversions are made, a traverse to other points and adjustments to traverses can be performed with ordinary ground traverses.

13. Reduction of Ground Distances to Grid Distances

In most situations, the modern survey total station measures a slope distance and vertical angle, which are then converted to a horizontal distance and vertical distance. This horizontal distance is then determined at the ground elevation of the instrument. The first step is to reduce this ground horizontal distance to a *geodetic distance*, also known as the *sea-level distance*. This conversion value is sometimes called the *elevation factor*. The geodetic distance is the distance measured along the ellipsoid chosen for the project.

In Fig. 11.4, R_e is the radius of the ellipsoid. This value can be calculated precisely for any point using the constants for the chosen ellipsoid. For most purposes, an average radius is acceptable. This average is 6 372 161.54 m (20,906,000 ft). H is the average elevation of the ground horizontal distance. N is the *geoid separation*, which is the distance between the surface of the geoid and the surface of the ellipsoid for any point. This value is not a constant and is generally avoided by NGS for the location of point P. It cannot be easily calculated by the engineer or surveyor. Within the continental United States, N is always a negative value. The formula (where L is the geodetic distance and D is the horizontal ground distance) for determining the geodetic distance is

$$L = D \left(\frac{R_e}{R_e + N + H} \right)$$

Note that L/D is the *elevation factor*.

Once the geodetic distance is determined, the grid distance can be calculated by multiplying the geodetic distance by the *scale factor*, SF. The scale factor is similar to the elevation factor in that it reduces the distance from one surface to another—in this case from the ellipsoid to the grid plane.

The scale factor is determined from the projection tables and is tabulated for every minute of latitude. In Table 11.2, the scale factor is given in the last column. If the scale factor is less than 1, the grid plane will be below the ellipsoid; if the scale factor is greater than 1, the grid plane will be above the ellipsoid. The exact scale factor for a given latitude must be calculated by proportioning the values given in the projection tables, similar to determining the value of R. The grid distance is calculated by

$$L_g = L(\text{SF})$$

Example 11.3

The ground slope distance from point A to point B is 3200.946 m, and the zenith angle (Z) is $87°46'55''$. The elevation of point A is 1000.000 m, and its latitude is $33°30'42''$ N. The geoid separation is -30.987 m.

Determine the grid distance between points A and B.

Solution

First the ground slope distance (SD) must be reduced to a ground horizontal distance (HD).

$$\begin{aligned} \text{HD} &= (\text{SD})(\sin Z) \\ &= (3200.946 \text{ m})(\sin 87°46'55'') \\ &= 3198.548 \text{ m} \end{aligned}$$

Next, multiply this horizontal distance by the elevation factor (EF) to determine the geodetic distance, L.

$$\begin{aligned} \text{EF} &= \frac{R_e}{R_e + N + H} \\ &= \frac{6\,372\,161.54}{6\,372\,161.54 + (-30.987) + 1000.000} \\ &= 0.99984795 \end{aligned}$$

$$\begin{aligned} L &= (\text{HD})(\text{EF}) \\ &= (3198.548 \text{ m})(0.99984795) \\ &= 3198.062 \text{ m} \end{aligned}$$

The SF must now be determined for the given latitude.

$$\text{latitude} = 33°30'42'' \text{ N}$$
$$\text{SF for } 33°30' = 0.99995833$$
$$\text{SF for } 33°31' = 0.99995921$$

$0.99995921 - 0.99995833 = 0.00000088$, which equals the change in SF for 1 min of latitude.

The SF for latitude $33°30'42''$ is determined by proportions.

$$\begin{aligned} x &= \frac{(0.00000088)(42)}{60} \\ &= 0.00000062 \end{aligned}$$

Adding this value to the SF for $33°30'42''$ gives the SF for point A.

$$\begin{aligned} \text{SF} &= 0.99995833 + 0.00000062 \\ &= 0.99995895 \end{aligned}$$

The grid distance is now determined by multiplying the geodetic distance by the SF.

$$\begin{aligned} \text{grid distance} &= (3198.062 \text{ m})(0.99995895) \\ &= 3197.931 \text{ m} \end{aligned}$$

Each distance in the traverse must be reduced in a similar manner before the traverse can be calculated. In CCS27, the reduction of ground distances to grid distances is the same except there is no geoid separation value.

14. Reduction of True/Geodetic Azimuth to Grid Azimuth

Before a direction can be reduced to grid azimuth, the type of azimuth must be known. If the azimuth is already grid, then no further reduction is necessary. This situation occurs when two grid control points are used for the basis of bearings of a traverse.

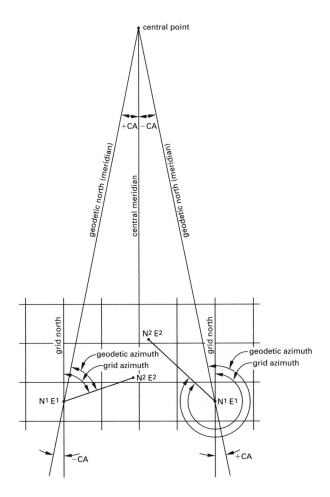

Figure 11.8 *Relationship Between Geodetic and Grid Azimuths*

If the current basis of bearings is magnetic north, the azimuth must first be reduced to a true azimuth by applying the magnetic declination. A *true azimuth* (TruA), also known as an *astronomic azimuth*, is based on celestial observation. Many recorded maps in California are based on true north.

As shown in Fig. 11.8, a true azimuth must be converted to a *geodetic azimuth*. Since true azimuths are based on the geoid and geodetic azimuths are based on the referenced ellipsoid, a correction must be made because of the difference in the direction of vertical of the two surfaces. This correction is the *Laplace correction*, another value that is usually supplied by NGS. It is generally a small value, rarely more that approximately 20 seconds. The Laplace correction is always added to the true azimuth. The geodetic azimuth is converted to a grid azimuth by applying the convergence angle (t is the grid azimuth, GeoA is the geodetic azimuth, and CA is the convergence angle).

$$t = \text{GeoA} - \text{CA}$$

Example 11.4

The true azimuth from point A to point B is $92°13'14''$. The Laplace correction for point A is $-14''$, and the convergence angle is $-0°23'44.6''$.

Determine the grid azimuth from point A to point B.

Solution

First, convert the true or astronomic azimuth to a geodetic azimuth.

$$\begin{aligned}\text{GeoA} &= \text{TruA} + \text{Laplace correction}\\ &= 92°13'14'' + (-14'')\\ &= 92°13'00''\end{aligned}$$

Next, convert the geodetic azimuth to a grid azimuth.

$$\begin{aligned}t &= \text{GeoA} - \text{CA}\\ &= 92°13'00'' - (-0°23'44.6'')\\ &= 92°36'44.6''\end{aligned}$$

15. Conversion of Coordinates from One Zone to Another

To convert grid coordinates in the overlap area of zones from one zone of the SPC system to another, convert the grid coordinates of the original zone using the constants and tables for that zone to geodetic coordinates. Then, using the constants and tables for the new zone, convert the geodetic coordinates to grid coordinates in the new zone.

Practice Problems

1. What are the advantages of placing surveys on the California Coordinate System?

2. What corrections must be made to measured slope distances prior to computing state plane coordinates?

3. How are surveys extended from one state plane zone to another, or from one state to another?

4. Develop a table of elevation factors for ground elevations ranging from sea level to 2000 m above sea level using 200 m increments.

5. Given:

$$R_b = 9\,836\,091.7896 \text{ m}$$
$$N_b = 500\,000 \text{ m}$$
$$E_o = 2\,000\,000 \text{ m}$$
$$CM = 116°15'$$
$$l = 0.549517575763$$
$$R \text{ for } 33°30' = 9\,688\,220.9733$$
$$R \text{ for } 33°31' = 9\,686\,372.4906$$
$$\text{tabular difference} = 30.80804 \text{ m}$$
$$\text{latitude of station A} = 33°30'17.12346''$$
$$\text{longitude of station A} = 118°02'13.98765''$$

Determine the grid coordinates of station A.

6. A survey party occupies station C having published coordinates of 539 034.888 m north and 1 977 009.714 m east. They measure a ground slope distance of 3908.789 m to station D. The zenith angle is $92°12'30''$. They determine the grid azimuth to be $120°15'00''$. The elevation of the station C is 500.000 m. The scale factor for station C is 0.99998765. What are the coordinates of station D?

Solutions

1. ■ It is a method for determining the plane coordinates (northing and easting) of a point from its geodetic coordinates (latitude and longitude).

■ It allows the use of plane survey techniques and calculations over large areas without introducing significant errors.

■ It establishes a single reference system for all surveys in an area. With the advent of Geographical Information Systems (GIS) where data from a large area must be collected and referenced to a common datum, CCS is a must.

■ It is a reference system that allows for easy retracement of survey work done previously on the CCS. It also creates a uniform computational base.

2. Measured slope distances must be corrected prior to use in determining state plane coordinates. Corrections include:

■ reducing the measured slope distance to a horizontal distance at a known elevation

■ correcting the horizontal distance to geodetic (sea level) distance using the elevation factor

■ correcting the geodetic distance to grid distance by applying the scale factor

The grid distance can then be used in calculations involving state plane coordinates.

3. The common factor in any state plane coordinate system is the geodetic coordinates (latitude and longitude). State plane coordinates in any zone can be transformed to geodetic coordinates and then transformed to state plane coordinates in another zone. This principle also applies to states that use the transverse mercador projection. Remember that CCS83 coordinates in one zone can only be transformed to CCS83 coordinates in another zone; CCS83 coordinates cannot be transformed to CCS27 coordinates.

4. Using 6 372 160 m as the radius of the earth and ignoring the geoid height values, the elevation factors are:

elevation	elevation factor
sea level	1.00000000
200 m	0.99996861
400 m	0.99993723
600 m	0.99990585
800 m	0.99987447
1000 m	0.99984309
1200 m	0.99981172
1400 m	0.99978034
1600 m	0.99974897
1800 m	0.99971760
2000 m	0.99968623

5. Determine the grid coordinates for the given data.

First, determine R for latitude $33°30'17.12346''$ N.

$$\text{distance} = (17.12346)(30.80804)$$
$$= 527.5402 \text{ m}$$
$$R = 9\,688\,220.9733 \text{ m} - 527.5402 \text{ m}$$
$$= 9\,687\,693.4331 \text{ m}$$

Next, determine CA.

$$\begin{aligned}\text{longitudinal} \\ \text{difference}\end{aligned} = 116°15' - 118°02'13.98765''$$
$$= -1°47'13.98765''$$
$$= -1.7872187917°$$
$$CA = (-1.7872187917°)(0.5495\,1757\,5763)$$
$$= -0.9821081378°$$
$$= -0°58'55.5893''$$

Then determine $R\cos CA$ and $R\sin CA$.

$$R\cos CA = 9\,686\,270.2751 \text{ m}$$
$$R\sin CA = -166\,048.8213 \text{ m}$$

Finally, determine the northing and easting of sta A.

$$\text{northing} = 9\,836\,091.7896 \text{ m} - 9\,686\,270.2751 \text{ m}$$
$$+ 500,000 \text{ m}$$
$$= 649\,821.515 \text{ m}$$
$$\text{easting} = 2\,000\,000 \text{ m} + (-166\,048.8215 \text{ m})$$
$$= 1\,833\,951.179 \text{ m}$$

6. The ground slope distance is 3908.789 m.

$$\begin{aligned}\text{horizontal distance} \atop \text{(at 500 m elev)} &= (3908.789 \text{ m})(\sin 92°12'30'')\\ &= 3905.8860 \text{ m}\end{aligned}$$

$$\begin{aligned}\text{geodetic distance} &= L\\ &= (3905.8860 \text{ m})\\ &\quad \times \left(\frac{6\,372\,161.54 \text{ m}}{6\,372\,161.54 \text{ m} + 500 \text{ m}}\right)\\ &= 3905.5796 \text{ m}\end{aligned}$$

$$\left[\begin{array}{l}\text{Since a geoid height value is not given,}\\ \text{use 0 for } N; \text{ do not assume a value for } N.\end{array}\right]$$

$$\begin{aligned}L_g = L(\text{SF}) &= (3905.5796)(0.99998764)\\ &= 3905.5313 \text{ m}\end{aligned}$$

Traverse station C to station D.

sta	azimuth	distance (m)	northing—lat (m)	easting—dep (m)
C			539,034.888	1,977,009.714
	120°15'00''	3905.5313		
D			537,067.383	1,980,383.451

Astronomical Observations 12

1. Introduction

People of ancient times were very much aware of their dependence on the cycles of earth and sky. The sun's warmth and radiance, the waxing and waning of the moon, and the regular movements of the stars and planets inspired their reverence for and belief that the heavenly bodies were gods who controlled the universe.

Temple priests observed the movements of these deities and were responsible for noting other celestial events and interpreting them. Storytellers told colorful stories about these gods and their interactions with humankind.

A great deal of astronomical knowledge was accumulated and passed on orally through these myths. Only the Babylonians and Egyptians recorded their findings for future civilizations.

The Greeks, from 700 B.C. to A.D. 200, first gave serious thought to the size and shape of the earth, and they were the first to conclude that the earth is a sphere. Aristotle noticed that the positions of the stars change and that during an eclipse the shadow of the earth on the moon is curved. Eratosthenes calculated the circumference of the earth to be 25,000 mi. Ptolemy, the great cartographer, devised a system for locating a point on the earth using a system similar to present-day latitude and longitude.

After Ptolemy, civilization slipped into the Dark Ages. For more than a thousand years the theory that the earth is a sphere was forgotten.

In the fifteenth century, the beginning of the European Renaissance, there was renewed interest in astronomy. Copernicus concluded that the earth revolved around the sun, directly contradicting the beliefs of the Catholic church. In the seventeenth century, Kepler discovered that the orbits of the planets around the sun are elliptical, with the sun at one focus of the ellipse. He also discovered that a line joining a planet and the sun sweeps out equal areas of space in equal time, proving that a planet moves fastest when nearest the sun and slowest when farthest away.

Galileo, the first to appreciate the significance of the telescope, accepted Copernicus' theories and was tried by the Inquisition for his heretical ideas.

2. Earth

The earth rotates from west to east on its polar axis and revolves about the sun in an elliptical orbit with the sun at one focus of the ellipse. It completes one revolution in a period of 365.2564 days. The inclination of the earth ($23^1/_2°$ with the perpendicular to the orbital plane), combined with its revolution around the sun, causes the lengths of day and night to change as well as the seasons (Fig. 12.1).

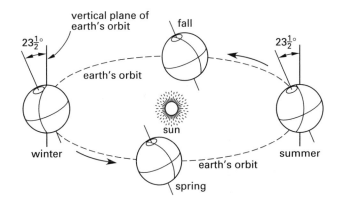

Figure 12.1 *Revolution of the Earth About the Sun*

On March 21 and September 23, the light from the sun reaches from one pole to the other. On these dates, shown as spring and fall in Fig. 12.1, the sun is directly overhead at the equator. On June 21, shown as summer, the sun is directly overhead at points $23^1/_2°$ above (north of) the equator. On December 22, shown as winter, the sun is overhead at points $23^1/_2°$ below (south of) the equator.

A line around the earth, parallel to and $23^1/_2°$ north of the equator, where the sun is directly overhead at its northernmost position, is known as the *Tropic of Cancer*. A line around the earth, parallel to and $23^1/_2°$ south of the equator, where the sun is directly overhead at its southernmost position, is known as the *Tropic of Capricorn*.

March 21 and September 23, when the sun crosses the equator and day and night are everywhere of equal length, are known as *equinoxes* (equal nights). The vernal equinox (March 21) marks the beginning of spring, and the autumnal equinox (September 23) marks the beginning of fall. June 21, the longest day of the year in the northern hemisphere, is known as the *summer solstice*. December 22, the shortest day of the year in the northern hemisphere, is known as the *winter solstice*.

In the southern hemisphere, the seasons are opposite those in the northern hemisphere.

3. Geodetic North or Geodetic Azimuth

The direction of a line is determined by the horizontal angle between the line and a reference line, usually true (geodetic) north. Determining the true azimuth of a line involves observations on a celestial body, such as the sun or another star. The star usually used in the northern hemisphere is Polaris, the North Star. It is selected because it is very near true north from any point north of the equator. In making observations on celestial bodies, the surveyor is not interested in the distance to these bodies from the earth, but merely in their angular positions from the observation point.

In addition to determining true azimuth, the surveyor can determine the latitude and longitude of his or her position by making observations on celestial bodies.

Observations on celestial bodies include making angular measurements, both horizontal and vertical. Determining true azimuth requires using an ephemeris (see Sec. 5).

4. Practical Astronomy

Since ancient times, practical astronomy has regarded all celestial bodies as being fixed onto a sphere of infinite radius whose center is the earth's center. (However, stars are not the same distance from the earth, and they are much farther away than they appear to be.) This sphere of infinite radius is called the *celestial sphere* (Fig. 12.2). To the observer, this sphere appears to be rotating about an axis, but it is not; the rotation of the earth causes the stars to appear to be in rotation. Practical astronomy assumes that the earth is stationary and that the celestial sphere revolves about the earth from west to east.

The celestial sphere is assumed to rotate about the celestial axis, which is a prolongation of the earth's polar axis. The north and south poles become the *north celestial pole* and the *south celestial pole*.

The plane of the earth's equator extended to the celestial sphere becomes the *celestial equator*.

Comparable with meridians of longitude of the earth are the *hour circles* of the celestial sphere, all of which converge at the celestial poles. They are also known as *celestial meridians*, as shown in Figs. 12.2 and 12.3.

Comparable with the parallels of latitude of the earth are the *parallels of declination* of the celestial sphere. They measure the angular distance from the celestial equator to the north and south celestial poles (Fig. 12.5).

The observer's position on the earth is located by latitude and longitude. If the observer's plumb line is extended upward to the celestial sphere, the point of intersection with the celestial sphere is the observer's *zenith*. If the plumb line is extended downward to the celestial sphere, the intersection is the observer's *nadir*.

The relative positions of an observer on the earth and on the celestial sphere are shown in Fig. 12.3. The angle at the center of the earth that measures latitude is the same for the earth and for the celestial sphere. The angle that measures longitude is also the same for the earth and for the celestial sphere.

In Fig. 12.3, the observer's position on earth is latitude 35° N and longitude 98° W. The arc distance (in degrees) between the zenith of the observer and the celestial equator is 35° N. The arc distance (in degrees) along the celestial equator between the planes of the Greenwich meridian and the meridian of the observer, extended to the celestial sphere is 98° W. This arc distance is also the angle at the celestial north pole between the planes of the two meridians.

By using latitude and longitude and projecting it to the celestial sphere, the position of the observer's instrument is fixed at a point on the celestial sphere.

To identify the location of a celestial body on the celestial sphere, a coordinate system similar to latitude and longitude was devised (Fig. 12.5). The celestial coordinates are called declination and right ascension. Using declination and right ascension as coordinates, the location of any celestial body on the celestial sphere can be determined with respect to the observer's zenith on the celestial sphere and to the north celestial pole.

Declination is the star's angular distance north or south of its celestial equator measured along the hour circle of the star (Fig. 12.5). North declination is plus (+); south declination is minus (−). Declination corresponds to latitude on the earth.

Right ascension is the arc distance measured eastward along the celestial equator from the vernal equinox to the hour circle of the star. It may be measured in degrees, minutes, and seconds of arc or in hours, minutes, and seconds of time (Fig. 12.5). Right ascension corresponds with longitude on the earth.

As with latitude and longitude, a celestial coordinate system requires points of origin, such as the equator for latitude and the Greenwich meridian for longitude. The celestial coordinate system uses the celestial equator, the ecliptic, and the vernal equinox in locating points of origin.

Because of the tilt of the earth as it follows its orbit around the sun, the sun traces a path called the *ecliptic* on the celestial sphere. The path of the sun moves from the southern hemisphere of the celestial sphere to the northern hemisphere and back (Fig. 12.4).

The point where the sun crosses the celestial equator on its movement each year from south to north along the ecliptic is known as the *vernal equinox* (Fig. 12.4). Astronomers designated the vernal equinox as the point of reference for right ascension.

The vernal equinox is a point on the celestial sphere of infinite distance from the earth. Its location in time, relative to the Greenwich meridian, is known.

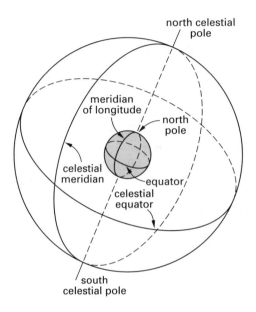

Figure 12.2 *Celestial Sphere*

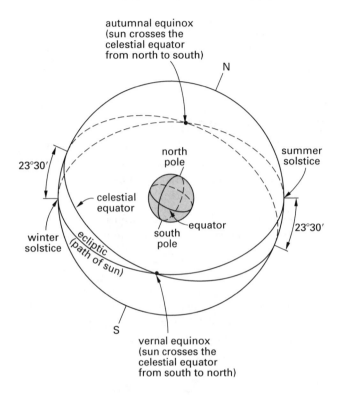

Figure 12.4 *Ecliptic and Vernal Equinox*

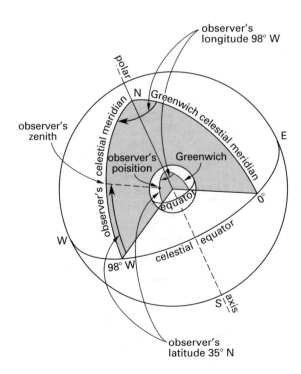

Figure 12.3 *Relation Between Earth and Celestial Sphere*

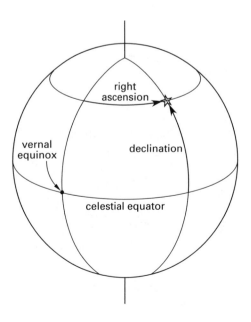

Figure 12.5 *Celestial Coordinates*

The point where the sky and earth meet, as seen by an observer on earth, is known as the *horizon*. The horizon for any place on the earth's surface is the great circle formed on the celestial sphere by the extension of the plane of the observer's horizon. In practical astronomy, the horizon is the plane tangent to the earth at the observer's position, perpendicular to the plumb line and extended to the celestial sphere. It is used as a reference for determining the altitude of a celestial body.

The *altitude* (h) of a celestial body is the angular distance measured from the horizon to a celestial body. It is the vertical angle measured by the observer from the horizon to the body (Fig. 12.9).

5. Astronomical Triangle

In order to determine azimuth or latitude and longitude, the surveyor needs to be able to solve a spherical triangle on the celestial sphere known as the *astronomical triangle* or the celestial triangle *PZS*.

The vertices of the *PZS* triangle are the north celestial pole (P), the observer's zenith (Z), and the position of the star or the sun (S), as shown in Fig. 12.6.

The sides of the triangle are arcs of great circles on the celestial sphere that pass through any two of the vertices, measured in degrees or hours. The angular value of each side is determined by the angle that the side subtends on the earth (Fig. 12.7).

The three sides of the *PZS* triangle are known as (1) the polar distance, (2) the coaltitude, and (3) the colatitude.

(1) The *polar distance* is the side *PS*, determined from the declination of the star or the sun. (Recall that declination is defined as the angular distance from a celestial body to the celestial equator.) When the celestial body lies north of the celestial equator, the declination has a positive sign; when it lies south of the celestial equator, it has a minus sign. The polar distance is determined algebraically by subtracting the declination of the celestial body from 90°. In Fig. 12.8, the polar distance is $90° - (-20°) = 110°$. Observations of stars south of the equator are seldom made from the northern hemisphere.

(2) The *coaltitude* is the side *SZ*, the arc distance from the celestial body to the observer's zenith. It is determined by subtracting the observed altitude h of the celestial body (corrected for refraction and parallax) from 90°. *Altitude* is the vertical angle measured from the observer's horizon to the celestial body.

(3) The *colatitude* is the side *PZ*. It is the distance from the celestial north pole to the zenith. It is determined by subtracting the latitude of the observer from 90° (Fig. 12.10).

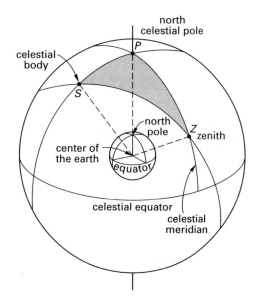

Figure 12.6 *PZS Triangle*

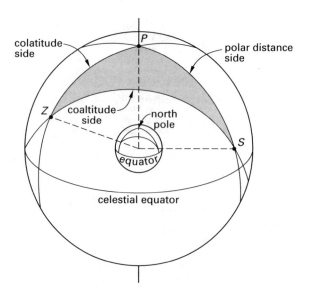

Figure 12.7 *Three Sides of the PZS Triangle*

The three interior angles of the *PZS* triangle are known as (1) the parallactic angle, (2) the azimuth or zenith angle, and (3) the angle *t* (Fig. 12.11).

(1) The *parallactic angle* (angle *S*) is formed by the polar distance side and the coaltitude side.

(2) The *azimuth angle* (angle *Z*) is formed by the coaltitude side and the colatitude side. It is used to find the azimuth from the observer to the celestial body. When the celestial body is in the east, the azimuth angle is equal to the true azimuth. When the celestial body is in the west, true azimuth equals 360° minus the azimuth angle.

(3) The angle at the pole *P* formed by the colatitude side and the polar distance side is known as the angle *t*, or the angle *P*.

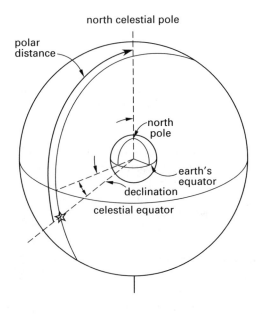

Figure 12.8 *Polar Distance Side of the PZS Triangle*

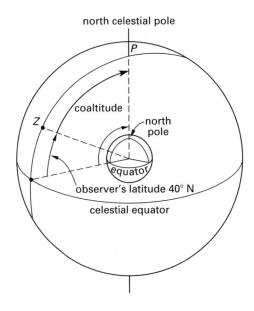

Figure 12.10 *Colatitude Side of the PZS Triangle*

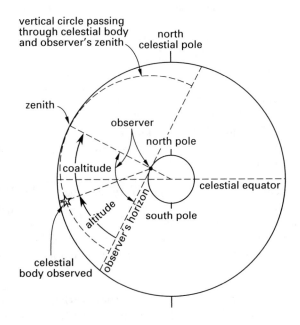

Figure 12.9 *Coaltitude Side of the PZS Triangle*

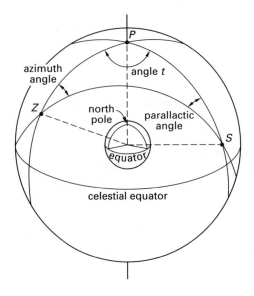

Figure 12.11 *Interior Angles of the PZS Triangle*

If any three elements of the *PZS* triangle are known, the other elements can be found by spherical trigonometry. However, each astronomical triangle is changing constantly because of the apparent rotation of the celestial sphere. The observer, then, must know the position of the *PZS* triangle at the time of his observation. Information concerning the position of the celestial bodies can be found in an ephemeris. (An *ephemeris* is similar to an almanac. It contains tables showing the positions of celestial bodies on certain dates in sequence. Typically, ephemerides are published by manufacturers of surveying instruments.)

6. Time

Because all celestial bodies are in constant apparent motion with respect to the observer, it is extremely important to know the precise time of an observation on a celestial body.

In practical astronomy, there are two categories of time: solar time and sidereal time. Both categories of time are based on the rotation of the earth with respect to a standard reference line.

Because the earth revolves around the sun in the plane of its orbit once each year, the reference line to the sun is changing constantly, and the length of one solar day is not the true time of one rotation of the earth.

In practical astronomy, the true time of one rotation of the earth, which is known as the *sidereal day*, is based on one rotation with respect to the vernal equinox. One solar day is 3 minutes 56 seconds longer than one sidereal day.

Explaining the difference between the solar day and the sidereal day requires temporarily abandoning the theory that the earth is stationary and that the celestial sphere is revolving and returning to the true condition that the earth revolves around the sun.

The earth completes one revolution around the sun in 365.2564 days, although one year is 365 calendar days. In Fig. 12.12, an observer at 0 on the earth on March 21 (the vernal equinox) would find the sun directly overhead at noon. From March 21 until the summer solstice, the observer would find that the sun is not directly overhead at noon, but advances about 1° north per day. This motion of the sun makes the intervals between the sun's transits (crossings) of the observer's meridian greater by about 3 minutes 56 seconds than the interval between transits of the vernal equinox of the observer's meridian. Therefore, the solar day is about 3 minutes 56 seconds longer than the sidereal day.

Because one apparent rotation of the celestial sphere is completed in one sidereal day, a star rises at nearly the

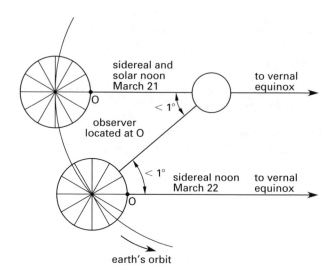

Figure 12.12 *Difference Between Solar Day and Sidereal Day*

same sidereal time throughout the year. On solar time, it rises about 4 minutes earlier from night to night, or 2 hours earlier from month to month. Thus, observed at the same hour night by night, the stars seem to move slowly westward across the sky as the year progresses.

An apparent solar day is the interval between two successive transits (crossings) of the sun over the same meridian. Because of the earth's tilt and the variation in the earth's velocity about the sun, the interval between two transits of the sun over the same meridian varies from day to day. This makes it impossible to use the variable *day* as a basis for accurate time. Therefore, a fictitious, or mean, sun was devised that is imagined to move at a uniform rate in its apparent path around the earth. It makes one apparent revolution around the earth in one year, the same as the actual sun. The average apparent solar day was used as a basis for the mean day. The time indicated by the position of the actual sun is called *apparent solar time*.

The difference between mean solar time and apparent solar time is called the *equation of time* (EOT). It varies from minus 14 minutes to plus 6 minutes (Fig. 12.13). The value of EOT for any day can be found in an ephemeris.

In mean solar time, the length of the year is divided into 365.2422 mean solar days. Because the mean sun appears to revolve around the earth every 24 hours of mean time, the apparent rate of movement of the mean sun is 15° of arc, or of longitude, per hour ($360 \div 24 = 15$).

In the system of latitude and longitude on the earth, the zero reference for latitude is the equator; the zero reference for longitude is the meridian that passes through

Greenwich, England (the prime meridian, or 0° longitude). Using the Greenwich meridian as a basis for reference, the time at a point 15° west of the Greenwich meridian is 1 hour earlier than the time at the Greenwich meridian because the sun passes the Greenwich meridian 1 hour before it crosses the meridian lying 15° to the west. The opposite is true along the meridian lying 15° to the east, where time is 1 hour later, because the sun crosses this meridian 1 hour before it arrives at the Greenwich meridian. Therefore, the difference in local time between two places equals their difference in longitude (Fig. 12.14).

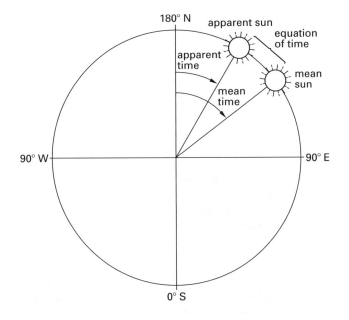

Figure 12.13 *Equation of Time*

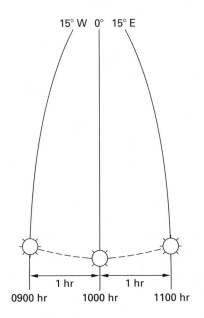

Figure 12.14 *Apparent Motion of the Sun*

Because the mean solar day has been divided into 24 equal units of time (hours), there are 24 time zones, each 15° wide, around the earth. Using the Greenwich meridian as the central meridian of a time zone and as the zero reference for the computation of time zones, each 15° zone extends 7½° east and west of the zone's central meridian (Fig. 12.15).

The central meridian of each time zone, east or west of Greenwich, is a multiple of 15°. For example, the time zone of the 90° meridian extends from 82°30′ to 97°30′. Each 15° meridian or multiple thereof east or west of the Greenwich meridian is called a *standard time meridian*. Four of these meridians (75°, 90°, 105°, 120°) cross the United States (Fig. 12.16).

Standard time zones in the United States are named Eastern, Central, Mountain, and Pacific. Standard time zone boundaries often run along state boundaries so that time is the same over a single state.

Standard time in any time zone is referred to as *local mean time*. It is clock time where the observer's position is located. It does not take into account daylight savings time.

Rather than establishing a reference meridian for measuring time in each time zone, it was decided to establish a single reference line for all parts on the earth. Standard time zone Z, which uses the Greenwich meridian as its basic time meridian, was chosen for computing data pertaining to mean solar time. Greenwich standard time is also Greenwich mean time (GMT). It is referred to as *Universal Time* in an ephemeris. Central Standard Time (CST) is standard time zone S.

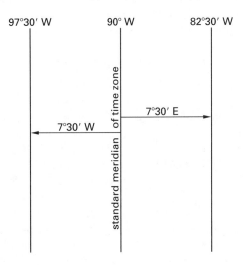

Figure 12.15 *Time Zone Boundaries*

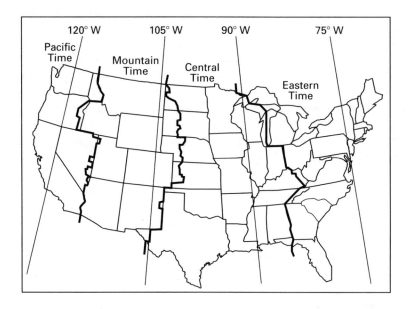

Figure 12.16 *United States Standard Time Zones*

Time can be expressed as the reading of the standard 24-hour clock at the Greenwich Observatory at the moment an observation is made on a celestial body; therefore, it is the same time throughout the world. Because the observer's watch is usually set to the local standard time (local mean time, or LMT), a conversion must be made from LMT to GMT. Data in an ephemeris are based on the Greenwich meridian and zero-hour Greenwich mean time.

To convert LMT to GMT when the observer is located in west longitude, divide the value of the central meridian of the time zone in degrees of longitude by 15°. This equals the time zone correction in hours. The difference in time between the standard time zone of the observer's position and GMT must be added to the LMT to arrive at the Greenwich mean time of observation (Fig. 12.18). If the result is greater than 24 hours, the first 24 hours are dropped and 1 day is added to obtain the Greenwich time and date. If the observer is located in east longitude, the difference is subtracted.

When observers sight the sun, it is obvious that they observe the apparent sun and not the mean sun on which their time is based. Therefore, they must convert mean time to apparent time, which is done by converting GMT at the point of observation to Greenwich apparent time (GAT). They first convert LMT to GMT by adding the time zone correction (Fig. 12.17) to LMT. In the Central time zone, 6 hours would be added. To obtain GAT, the equation of time is added to GMT for observations in west longitude and subtracted from GMT for east longitude. The equation is found in an ephemeris, using the date and time of observation. In summary, for west longitude,

local mean time
+ time zone correction
  = GMT
    + equation of time for zero hour
      (from ephemeris)
    + daily change (from ephemeris)
  = GAT

An *hour angle* is any great circle on the celestial sphere that passes through the celestial poles. It corresponds to a meridian on the earth.

The observer's meridian is the great circle on the celestial sphere that passes through the celestial poles and the observer's zenith.

The hour angle of a celestial body is the angle at the celestial poles between the plane of the meridian of the observer and the plane of the hour angle of the celestial body. In Fig. 12.11, it is shown as the angle P; it is also known as the angle t.

The *Greenwich hour angle* (GHA) of a celestial body is the time that has elapsed since the body crossed the Greenwich meridian (projected on the celestial sphere).

The *local hour angle* (LHA) of a celestial body is the angle measured along the plane of the celestial equator from the meridian of the observer (zenith) to the meridian of the celestial body (projected on the celestial sphere).

For west longitude, where λ is the longitude of the observer's position,

$$LHA = GHA - W\lambda \qquad 12.1$$

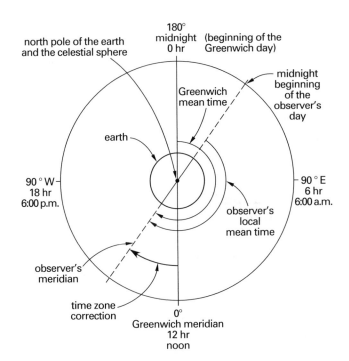

Figure 12.17 *Time Zone Correction*

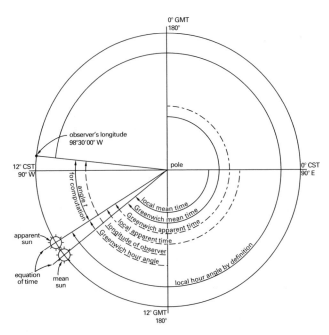

Figure 12.18 *Time; Hour Angle Relationship and West Longitude*

Example 12.1

The observer's longitude is $98°30'00''$ W, and the local mean time of observation is 09h 00m 00s CST, 24 April 68. Determine the angle t for the solution of the PZS triangle.

Solution

	09h	00m	00s	
+	06	00	00	time zone correction ($90°$ W $\div$ $15°$ = +6)
	15	00	00	GMT of observation
+		01	44	equation of time for zero-hour GMT*
+		00	07	equation of time for partial day*
	15	01	15	GAT of observation
−	12	00	00	GHA measured from noon
	03	01	51	GHA = $45°27'45''$
	$45°$	$27'$	$45''$	
+	$360°$			(if necessary)
	$405°$	$27'$	$45''$	
−	$98°$	$30'$	$00''$	longitude of observer
	$306°$	$57'$	$45''$	LHA of the sun

and $53°$ $02'$ $15''$ angle t $(360° - 306°57'45'')$

*from ephemeris

The following factors can be used in converting hours to degrees and degrees to hours.

24 h	=	$360°$		$360°$	=	24 h
1 h	=	$15°$		$1°$	=	4 m
1 m	=	$15'$		$1'$	=	4 s
1 s	=	$15''$		$1''$	=	0.067 s

As mentioned previously, Greenwich mean time, which is also Greenwich standard time, is referred to in an ephemeris as *Universal Time*, or UTC (Coordinated Universal Time). This time is broadcast by the radio station WWV of the National Bureau of Standards and can be received on receivers that are pretuned to WWV. Also, the time signals can be obtained by calling (303) 449-7111 at the caller's expense.

The time signals can be used to determine a more precise time called UT1. UT1 is obtained by adding a small correction called DUT (UT1 = UTC + DUT).

The DUT correction can be determined by listening carefully to the WWV time signal. Following a minute tone, there will be a number of double ticks. Each double tick represents a correction of 0.1 second and is positive for the first 7 seconds. Beginning with the ninth second, each double tick is a negative correction. For example, a voice on the radio will announce "fifteen hours thirty-six minutes." Just after this will be a minute tone followed by the double ticks. This occurs for each minute.

The DUT correction changes 0.1 second periodically, but not uniformly; it does not change rapidly. It may remain constant for a week or more in some instances.

Ephemerides published in 1988 are based on UT1 time for observations on the sun. Data needed for azimuth calculations are tabulated in an ephemeris for each day of the year for zero-hour Universal Time, so that it is possible for the day of the month at Greenwich to be one day later than the date of the observation. At

6:00 p.m. CST it is midnight at Greenwich, so for observations on Polaris after 6:00 p.m. CST, one day would be added to the local date to find the Greenwich date in order to enter the tables in an ephemeris. For most sun shots, Greenwich date and local date will be the same.

The *sidereal day* is defined by the time interval between successive passages of the vernal equinox over the upper meridian of a given location. The sidereal year is the interval of time required for the earth to orbit the sun and return to the same position in relation to the stars. Because the sidereal day is 3 minutes 56 seconds shorter than the solar day, this differential in time results in the sidereal year being one day longer than the solar year, or a total of 366.2422 sidereal days. And, because the vernal equinox is used as a reference point to mark the sidereal day, the sidereal time for any point at any instant is the number of hours, minutes, and seconds that have elapsed since the vernal equinox passed the meridian of the point.

The general steps for converting local mean time of observation to local hour angle (then to the interior angle t at the pole) from sidereal time are as follows.

step 1: Greenwich mean time of observation is determined the same way as solar time is determined.

step 2: Sidereal time for zero-hour GMT plus the correction for GMT (from an ephemeris) determines the Greenwich sidereal time of observation.

step 3: Greenwich sidereal time of observation minus the right ascension of the star (from an ephemeris) equals the Greenwich hour angle.

step 4: Greenwich hour angle plus the observer's longitude (if in east longitude) or minus the observer's longitude (if in west longitude) is the local hour angle of the star.

step 5: Angle t, the interior angle of the *PZS* triangle at the pole, equals the local hour angle when the star is in the east. The specific steps performed in determining the local hour angle = angle t are as follows.

corrected watch time
  + time zone correction
    = GMT
      + sidereal time for zero hour (from ephemeris)
      + correction for partial day (from ephemeris)
    = greenwich sidereal time
      − right ascension of star (from ephemeris)
    = GHA
      − W longitude (+ for E longitude)
    = LHA = angle t

In general, it can be stated that observations on the sun involve apparent solar time, while observations on the stars are based on sidereal time. The computations using either apparent solar time or sidereal time are similar in that they do nothing more than fix the location of both the celestial body and the observer in relation to the Greenwich meridian. Once a precise relationship has been established, it is a simple matter to complete the determination of azimuth to the celestial body.

7. Methods and Techniques of Determining Azimuth

There are two methods of determining azimuth by astronomical observations: the *altitude method* and the *hour angle method*. Both methods require a horizontal angle from an azimuth mark on the ground to the observed body (sun or star) in order to establish azimuth on the ground. The basic difference between the two methods is that the altitude method requires an accurate vertical angle measurement but does not require precise time; the hour angle method requires precise time but does not require a vertical angle measurement. For observations made to the sun, an accuracy of ±10 seconds of arc can be expected when using the hour angle method. Using the altitude method will provide an accuracy of ±1 minute of arc. For observations made to Polaris, an accuracy equal to the observer's ability in measuring the horizontal angle can be obtained when using the hour angle method. The altitude method can be used but is generally not used in Polaris observations. For stars other than Polaris that are listed in the ephemeris, the accuracy will be similar to those obtained in solar observations.

In both methods, it is extremely important that the instrument be leveled carefully. The vertical axis must be truly vertical; if it is not, the error caused by the inclination of the horizontal plane will not be eliminated by a reversal of the telescope between sights. Centering the plate level bubbles on the instrument makes the vertical axis of the instrument truly vertical, provided that the plate levels are in perfect adjustment.

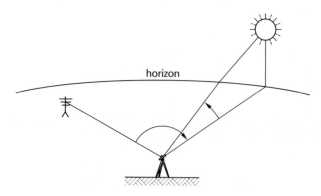

Figure 12.19 *Measuring Horizontal and Vertical Angles*

In the altitude method of determining azimuth, the *PZS* triangle is solved by using the three sides of the triangle. In addition to the horizontal angle from a ground point to the celestial body, three elements are necessary and must be determined: (1) the latitude for determining the colatitude side (colatitude equals 90° minus latitude), (2) the declination of the celestial body (angular distance from the celestial equator to the celestial body) for determining the polar distance side of the *PZS* triangle (polar distance equals 90° minus declination), and (3) the observed altitude (vertical angle) to the celestial body for determining the coaltitude side of the triangle (coaltitude equals 90° minus corrected altitude).

In the hour angle method of determining azimuth, the azimuth angle is determined from two sides and the included angle of the *PZS* triangle. The sides are the polar distance and the colatitude, as explained in the altitude method. The angle at the north celestial pole, the angle *t*, is determined as explained in Sec. 6.

The principal advantage of the altitude method is that precise time is not required. Until recent years, timepieces that make UT1 time possible in the field were not available. Because of this, the altitude method has been widely used.

A disadvantage of the altitude method is that a vertical angle is required for observation to both the sun and the stars, which makes it necessary to set both the horizontal and vertical cross-hairs tangent to the sun simultaneously. Also, measuring a vertical angle introduces the necessity of making corrections for parallax and refraction.

Advantages of the hour angle method counter the disadvantages of the altitude method. Bringing the vertical cross-hair tangent to the sun without concern for the horizontal cross-hair is much less difficult than achieving simultaneous tangency. Also, eliminating corrections for parallax and refraction (for sun sights) contributes to more accurate results.

Disadvantages of the hour angle method are the cost of timepieces and the additional training required to use them.

Both methods can be used for observation on either the sun or the stars. Both methods require the determination of the latitude and longitude of the point of observation. All in all, however, the hour angle method seems to be the preferred method.

The Lietz Ephemeris for 1988 offers an example problem for the altitude method.

In determining the azimuth of a line, the general equation is

$$\text{az line} = \text{az sun or star} + 360° \text{ angle right} \qquad 12.2$$

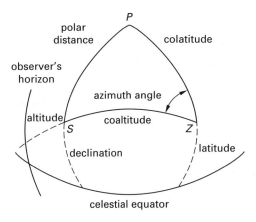

Figure 12.20 *Altitude Method*

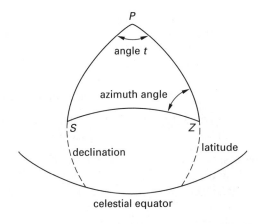

Figure 12.21 *Hour Angle Method*

Example 12.2

Find the azimuth of a line when the azimuth to the sun is 78°31′24.6″ and the angle right is 346°20′18.1″.

Solution

$$\text{az line} = 78°31′24.6″ + 360° - 346°20′18.1″$$
$$= 92°11′06.5″$$

8. Maps and Map Reading

Topographic maps of various scales can be purchased by mail for reasonable prices. Maps of areas east of the Mississippi River can be purchased from U.S. Geological Survey, 1200 South Eads Street, Arlington, VA 22202. Maps of areas west of the Mississippi River can be purchased from U.S. Geological Survey, Box 25286, Federal Center, Denver, CO 80225.

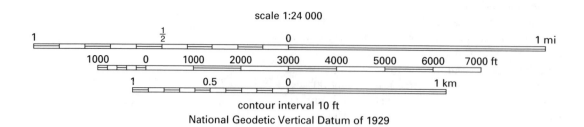

Figure 12.22 *Scale of the Elm Mott Map*

USGS quadrangle series maps cover areas bounded by parallels of latitude and meridians of longitude. Standard edition maps are produced at 1:24,000 scale in either 7.5 × 7.5- or 7.5 × 15-minute format. The 7.5-minute quadrangle map is satisfactory for determining latitude and longitude by scaling in determing azimuth.

The scale used in Fig. 12.22, 1:24,000, is convenient because 1 in = 24,000 in = 2,000 ft exactly.

The east side of the Elm Mott map (Fig. 12.23) is bounded by a line representing the meridian of 97°00′ west longitude; the west side is bounded by a line representing the meridian of 97°07′30″ west longitude. The south side is bounded by a line representing the parallel of 31°37′30″ north latitude; the north side is bounded by a line representing the parallel of 31°45′ north latitude. Thus, the quadrangle formed is 7.5 minutes on each side. The east and west lines of the map are marked by ticks at 2.5 minute intervals (31°40′ and 31°42′30″ north latitude), and the north and south lines are marked by ticks at 2.5 minute intervals (97°02′30″ and 97°05′ west longitude). Connecting corresponding ticks on the east and west lines with lines and connecting corresponding ticks on the north and south lines with lines divides the quadrant into nine 2.5′ × 2.5′ subquadrants. (The borders of the Elm Mott map are not shown to scale in Fig. 12.23.)

If the latitude and longitude of Monument BM No. 498 are needed, we first select the subquadrant that contains the monument and determine the lines of latitude and longitude that bound the subquadrant.

After the parallels and meridians for the subquadrant have been drawn (Fig. 12.24), the geographic interval (angular distance between two adjacent lines) must be determined. Examination of the tick marks gives the interval. On the 7.5′ quad map, the interval is 2′30″ = 150″. Any scale with 150 divisions may be used to find latitude and longitude. The 20 scale of an engineer's scale fits this requirement.

To find the longitude of the Monument BM No. 498, place the 0 mark of the twenty scale on meridian 97°00′ (east side because longitude increases from east to west) and the 15 mark of the scale on the meridian 97°02′30″ (west side) with the scale just above the monument. Then slide the scale downward, keeping the 0 and 15 marks on the lines, until the edge of the scale just touches the point of the monument. The reading on the scale will give the number of seconds west longitude from the meridian 97°00′. The scale reading is 66, so the longitude of the monument is 97°01′06″ west. Following the same procedure for latitude, the scale reading is 76, so the latitude of the monument is 31°43′46″ north.

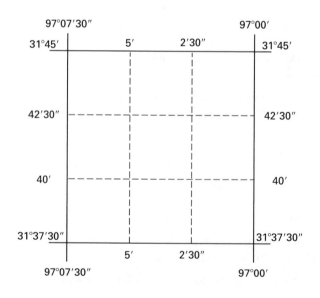

Figure 12.23 *Outline of the Elm Mott Map*

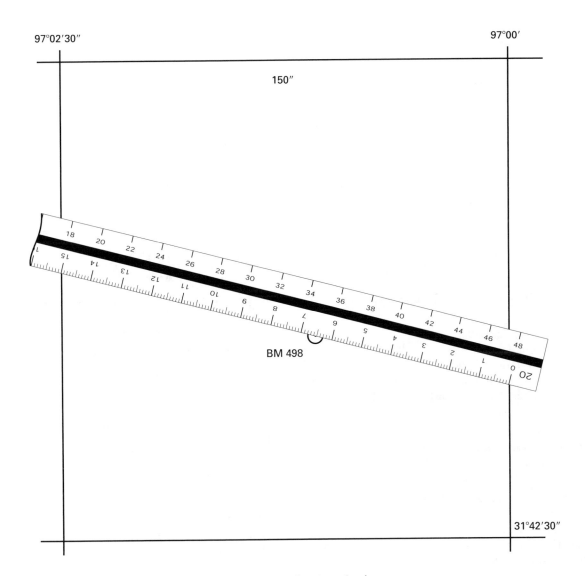

Figure 12.24 *Scaling Longitude*

9. Locating Polaris

Ancient astronomers identified some groups of stars with mythological characters, animals, and everyday objects, and named the groups accordingly. These groups of stars are called *constellations*. Two of the most well-known constellations are Ursa Major and Ursa Minor. Ursa Major translates to Great Bear, and it contains the prominent configuration of stars known as the Big Dipper. Ursa Minor is the Little Bear and contains the Little Dipper.

As mentioned in Sec. 6, because of the difference in solar time and sidereal time, a star apparently rises about 4 minutes earlier from night to night, or 2 hours earlier from month to month. Thus, at the same hour, night by night, a constellation seems to move slowly westward across the sky.

Because of the motion of the earth, the celestial sphere appears to rotate once a day, causing a constant movement of the constellations across the sky.

Stars that never set are said to be circumpolar. A star with declination greater than 90° minus an observer's latitude is circumpolar at that latitude. The stars in Ursa Major can be seen as far south as 30° south latitude, while the constellation Orion straddles the celestial equator and can be seen from anywhere in the world.

Polaris appears to move in a small, counterclockwise, circular orbit around the celestial north pole. Because Polaris stays so close to the north celestial pole, it is visible throughout most of the northern hemisphere. When the Polaris hour angle is 0 or 12 hours, the star is said to be in its upper or lower culmination. When the Polaris local hour angle is 6 or 18 hours, the star is said to be in its western or eastern elongation. When Polaris is near western or eastern elongation, it appears to move slowly and vertically.

Polaris is the brightest star in Ursa Minor, which is near Ursa Major and the constellation Cassiopeia. It is the end star of the three stars making up the handle of the Little Dipper. Polaris can also be identified with respect to the Big Dipper and Cassiopeia. The two stars forming the side of the bowl farthest from the handle of the Big Dipper are called the *pointer stars*. A line through the pointer stars toward Cassiopeia nearly passes through the celestial north pole. The distance from the nearest pointer star to Polaris is about five times the distance between the two pointer stars. Polaris and Cassiopeia are on the same side of the north celestial pole.

The vertical angle to Polaris is nearly the same as the latitude of the observer, depending on the position of Polaris in its orbit. In searching for Polaris, this angle can be set on the observing instrument. The telescope should first be focused on any bright star.

When the telescope is directed at Polaris, the observer will see two other stars nearby that are not visible to the naked eye. However, Polaris will be the only star visible when the cross-hairs are lighted.

The orbit of Polaris is shown in Fig. 12.25, and the constellations that serve to identify Polaris are shown in Fig. 12.26.

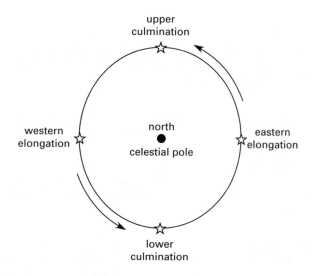

Figure 12.25 *Orbit of Polaris*

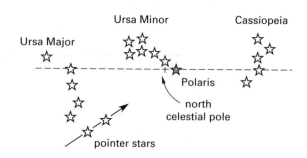

Figure 12.26 *Identification of Polaris*

10. Simpler Method of Determining Azimuth

In determining azimuth from astronomical observations, the work involved has been made easier by present-day ephemerides. Determining the Greenwich hour angle as in Ex. 12.1 has been simplified.

In 1988, ephemerides tabulate the GHA and the declination of the sun and Polaris at zero-hour Universal Time in degrees, minutes, and seconds for each day of the year, thus eliminating some of the computations in Ex. 12.1. Interpolation for an exact time of day is, of course, still necessary. The LHA is determined as before (LHA = GHA − Wλ).

Example 12.3

Given the following information, determine the azimuth of the sun.

> local date: June 7, 1988
> UT1 time: 14:32:26.2
> latitude: N 31°38′20″
> longitude: W 97°04′46″

Solution

Greenwich date and local date are the same.

$$\text{GHA} = 38°23′27.1″ \quad \text{[from ephemeris]}$$
$$\text{LHA} = \text{GHA} − \text{W}\lambda$$
$$= 38°23′27.1″ − 97°04′46″$$
$$= 301°18′41.1″$$
$$\text{decl} = 22°48′41.1″ \quad \text{[from ephemeris]}$$
$$\text{az to sun} = 84°17′96.5″ + \text{semidiameter}$$
$$\text{[from formula in ephemeris]}$$

$$84°17′9.5″ − 0°19′56.5″ = 83°57′13.0″$$
$$h_c = 37°40′40.1″$$

Table 12.1 Sample Page from Ephemeris

JUNE 1988

APPARENT PLACES OF THE SUN AND POLARIS FOR 0 HOUR UNIVERSAL TIME

DAY	GHA (SUN)	DECLINATION	EQ. OF TIME APPT-MEAN	SEMI-DIAM.	GHA (POLARIS)	DECLINATION	GREENWICH TRANSIT
	° ′ ″	° ′ ″	M S	′ ″	° ′ ″	° ′ ″	H M S
1 W	180 33 48.1	22 03 24.7	02 15.20	15 47.8	215 13 40.3	89 12 37.40	9 37 30.
2 TH	180 31 30.1	22 11 21.9	02 06.00	15 47.7	216 12 28.7	89 12 37.23	9 33 36.
3 F	180 29 06.2	22 18 56.0	01 56.42	15 47.6	217 11 18.5	89 12 37.09	9 29 41.
4 SA	180 26 36.8	22 26 06.7	01 46.45	15 47.4	218 10 10.1	89 12 36.95	9 25 46.
5 SU	180 24 01.9	22 32 54.0	01 36.13	15 47.3	219 09 03.2	89 12 36.81	9 21 51.
6 M	180 21 21.9	22 39 17.7	01 25.46	15 47.2	220 07 57.2	89 12 36.65	9 17 57.
7 TU	180 18 36.9	22 45 17.5	01 14.46	15 47.0	221 06 50.9	89 12 36.47	9 14 02.
8 W	180 15 47.3	22 50 53.4	01 03.15	15 46.9	222 05 43.4	89 12 36.26	9 10 07.
9 TH	180 12 53.3	22 56 05.2	00 51.55	15 46.8	223 04 34.0	89 12 36.04	9 06 12.
10 F	180 09 55.3	23 00 52.8	00 39.69	15 46.7	224 03 22.3	89 12 35.81	9 02 17.
11 SA	180 06 53.6	23 05 16.0	00 27.57	15 46.6	225 02 08.6	89 12 35.60	8 58 23.
12 SU	180 03 48.6	23 09 14.9	00 15.24	15 46.5	226 00 53.4	89 12 35.41	8 54 29.
13 M	180 00 40.8	23 12 49.2	00 02.72	15 46.4	226 59 37.7	89 12 35.25	8 50 34.
14 TU	179 57 30.4	23 15 59.0	−00 09.97	15 46.3	227 58 22.2	89 12 35.11	8 46 40.
15 W	179 54 18.0	23 18 44.1	−00 22.80	15 46.2	228 57 07.6	89 12 34.99	8 42 46.
16 TH	179 51 03.9	23 21 04.5	−00 35.74	15 46.1	229 55 54.2	89 12 34.89	8 38 51.
17 F	179 47 48.6	23 23 00.2	−00 48.76	14 46.0	230 54 42.1	89 12 34.80	8 34 57.
18 SA	179 44 32.4	23 24 31.1	−01 01.84	15 45.9	231 53 31.1	89 12 34.70	8 31 02.
29 SU	179 41 15.7	23 25 37.1	−01 14.95	15 45.9	232 52 20.9	89 12 34.60	8 27 07.
20 M	179 37 59.0	23 26 18.3	−01 28.07	15 45.8	233 51 10.8	89 12 34.48	8 23 13.
21 TU	179 34 42.6	23 26 34.7	−01 41.16	15 45.8	234 50 00.4	89 12 34.35	8 19 18.
22 W	179 31 26.9	23 26 26.3	−01 54.21	15 45.7	235 48 49.2	89 12 34.21	8 15 23.
23 TH	179 28 12.2	23 25 53.0	−02 07.18	15 45.7	236 47 36.8	89 12 34.06	8 11 29.
24 F	179 24 59.0	23 24 55.0	−02 20.07	15 45.6	237 46 22.8	89 12 33.91	8 07 34.
25 SA	179 21 47.5	23 23 32.3	−02 32.83	15 45.6	238 45 07.2	89 12 33.77	8 03 40.
26 SU	179 18 38.1	23 21 44.8	−02 45.46	15 45.5	239 43 50.0	89 12 33.65	7 59 46.
27 M	179 15 31.1	23 19 32.8	−02 57.93	15 45.5	240 42 31.7	89 12 33.54	7 55 52.
28 TU	179 12 26.8	23 16 56.3	−03 10.21	15 45.5	241 41 12.9	89 12 33.47	7 51 58.
29 W	179 09 25.5	23 13 55.3	−03 22.30	15 45.5	242 39 54.6	89 12 33.43	7 48 03.
30 TH	179 06 27.4	23 10 30.0	−03 34.17	15 45.4	243 38 37.6	89 12 33.41	7 44 09.

Used with permission from *1988 Celestial Observation Handbook and Ephemeris*, Dr. Richard L. Elgin, Dr. David R. Knowles, and Dr. Joseph Senne, published by the Lietz Company.

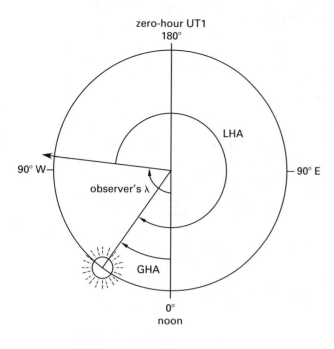

Figure 12.27 *Observation on Sun*

Example 12.4

Given the following information, determine the azimuth of Polaris.

> local date: June 7, 1988
> UTC time: 01:56:31
> latitude: N 31°38′20″
> longitude: W 97°04′46″

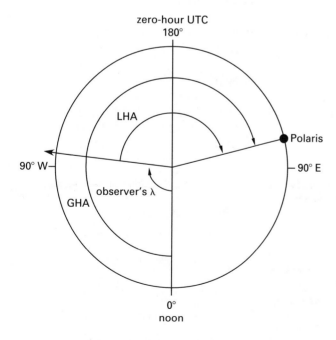

Figure 12.28 *Observation on Polaris*

Solution

Greenwich date is 1 day later than local date.

$$\text{GHA} = 251°18′13.8″ \quad \text{[from ephemeris]}$$
$$\text{LHA} = \text{GHA} - \text{W}\lambda$$
$$= 251°18′13.8″ - 97°04′46″$$
$$= 154°13′27.8″$$
$$\text{decl} = 89°12′36.2″ \quad \text{[from ephemeris]}$$
$$\text{az to Polaris} = 0°24′23.7″ \quad \text{[west of north]}$$

Practice Problems

1. On what day(s) of the year does light from the sun reach from one pole to another?

(A) June 21

(B) September 23

(C) December 22

(D) March 21

2. What is the type of azimuth obtained from a celestial observation?

(A) grid

(B) geodetic

(C) true

(D) assumed

3. What is the declination of a celestial body?

(A) the distance east or west of the Greenwich meridian on the celestial sphere

(B) the distance north or south of the poles on the celestial sphere

(C) the distance north or south of the equator on the celestial sphere

(D) the distance from the center of the earth to the celestial body being observed

4. What is the hour circle that contains a celestial body?

(A) the distance east or west of the Greenwich meridian on the celestial sphere

(B) the distance north or south of the poles on the celestial sphere

(C) the distance north or south of the equator on the celestial sphere

(D) the distance from the center of the earth to the celestial body being observed

5. Describe the vertices of the celestial triangle *PZS*.

6. Briefly describe the difference between an apparent solar day and a mean solar day.

7. Determine the azimuth of a line when the azimuth to the sun is $269°34'25''$ and the angle right is $87°56'57''$.

8. Given the following information, determine the azimuth of the sun. Use the data from Table 12.1.

$$\begin{aligned} &\text{local date:} &&\text{June 21, 1988} \\ &\text{UTC time:} &&16\text{:}42\text{:}21.3 \\ &\text{latitude:} &&33°42'38'' \text{ N} \\ &\text{longitude:} &&117°52'41'' \text{ W} \end{aligned}$$

Solutions

1. Answer (B)

   Answer (D)

2. Answer (C)

3. Answer (C)

4. Answer (A)

5. *P* is the north celestial pole, which is the intersection of the celestial sphere with the prolongation of the earth's polar axis.

Z is the observer's zenith, which is the intersection of the celestial sphere with the observer's plumb (vertical) line.

S is the position of the celestial body on the celestial sphere.

6. An apparent solar day is the interval between two successive crossings (transits) of the sun over the same meridian. Because of the tilt of the earth's axis and the elliptical orbit around the sun, the apparent solar day is generally not exactly 24 hours. This variation in the length of the day makes it difficult, if not impossible, to use apparent solar time as a basis for calculations.

Mean solar time is a theoretical time based on each day being exactly 24 hours. This time is kept by the bureau of standards and is broadcast over radio station WWV. This is the time that the data in the ephemeris is related to.

The difference between apparent solar time and mean solar time is called the *equation of time*. The equation of time is listed in the ephemeris.

7. $$\text{az} = \text{az of sun} + 360° - \text{angle right}$$

$$\begin{aligned} 269°34'25'' + 360° - 87°56'57'' &= 541°37'28'' \\ &= 181°37'28'' \\ &\qquad\text{[normalized]} \end{aligned}$$

If the angle right is less than the azimuth of the sun, the addition of $360°$ can be ignored.

8. $$\begin{aligned} &\text{GHA for June 21} = 179°34'42.6'' \ (0 \text{ hour}) \\ &\text{GHA for June 22} = 179°31'26.9'' \ (24 \text{ hour}) \end{aligned}$$

These values are for zero-hour (midnite) Greenwich time and must be interpolated for the time of observation.

Per the ephemeris:

$$\begin{aligned} \text{GHA} &= \text{GHA } 0^h + (\text{GHA } 24^h - \text{GHA } 0^h + 360°) \\ &\quad \times \left(\frac{\text{UT1}}{24} \right) \\ &= 179°34'42.6'' \\ &\quad + (179°31'26.9'' - 179°34'42.6'' + 360°) \\ &\quad \times \left(\frac{16\text{:}42\text{:}21.3}{24} \right) \\ &= 430°07'45.9'' \\ &= 70°07'45.9'' \end{aligned}$$

$$\begin{aligned} \text{LHA} &= \text{GHA} - \text{W}\lambda \\ &= 70°07'45.9'' - 117°52'41'' \\ &= -47°44'55.1'' \\ &= 312°15'04.9'' \end{aligned}$$

Declination:

$$\begin{aligned} &\text{decl for June 21} = 23°26'34.7'' \ (0^h) \\ &\text{decl for June 22} = 23°26'26.3'' \ (24^h) \end{aligned}$$

As with GHA, these values must be interpolated for the time of observation.

Per the ephemeris:

$$\begin{aligned} \text{decl} &= \text{decl } 0^h + (\text{decl } 24^h - \text{decl } 0^h) \left(\frac{\text{UT1}}{24} \right) \\ &\quad + (0.0000395)(\text{decl } 0^h) \sin(7.5 \text{ UT1}) \\ &= 23°26'34.7'' + (23°26'26.3'' - 23°26'34.7'') \\ &\quad \times \left(\frac{16\text{:}42\text{:}21.3}{24} \right) + (0.0000395) \\ &\quad \times (23°26'34.7'') \sin[(7.5)(16\text{:}42\text{:}21.3)] \\ &= 23°26'31.6'' \end{aligned}$$

The second term $(0.0000395\dots)$ added to the interpolated value changes the answer from a linear interpolation to a nonlinear two-point interpolation.

Azimuth to the sun:

$$\text{az} = \tan^{-1}\left(\frac{-\sin\left(\text{LHA}\right)}{\cos\theta\tan\delta - \sin\phi\cos\left(\text{LHA}\right)}\right)$$

$$[\text{from ephemeris}]$$

$$\phi = \text{latitude}$$
$$\delta = \text{declination}$$
$$\text{LHA} = \text{local hour angle}$$

$$\text{az} = \tan^{-1}\left(\frac{-\sin\left(312°15'04.9''\right)}{\begin{array}{c}\cos\left(33°42'38''\right)\tan\left(23°26'31.6''\right)\\ -\sin\left(33°42'38''\right)\cos\left(312°15'04.9''\right)\end{array}}\right)$$

$$= -89°02'05.1'' + 180° \quad \left[\begin{array}{l}\text{normalization}\\ \text{per ephemeris}\end{array}\right]$$

$$= 90°57'54.9''$$

United States Public Lands System

13

1. Introduction

The United States Public Lands Survey System (USPLSS) was conceived in 1785 by Thomas Jefferson as a method for dividing and describing land for sale to the settlers moving west from the original 13 states.

2. Section

The main subdivision of land in the system is the *section*. A *regular section* is approximately one mile square. The unit of measurement adopted by the government is the *Gunter's chain*. The chain is 66 ft long and contains 100 links. There are 80 chains to the mile, therefore the section is 80 chains on a side and contains 640 acres. Sections may also be fractional (less than 640 acres). A section may be made fractional for many reasons. Its boundaries may be interrupted by senior grant or reservation lines, by bodies of water, or by property lines. The section may contain bodies of water, the surface areas of which are deducted from the area of the section. Title to these bodies of water is vested in the state in which the land is located; therefore, the area was not included when the land passed from public to private ownership.

***Table 13.1** Units of Measurement*

$$1 \text{ chain} = 66 \text{ feet} = 100 \text{ links}$$
$$1 \text{ link} = 0.66 \text{ feet} = 7.92 \text{ inches}$$
$$80 \text{ chains} = 5280 \text{ feet} = 1 \text{ mile}$$
$$10 \text{ square chains} = 1 \text{ acre} = 43{,}560 \text{ ft}^2$$
$$1 \text{ section} = 80 \text{ chains} \times 80 \text{ chains}$$
$$= 6400 \text{ square chains}$$
$$= 640 \text{ acres}$$

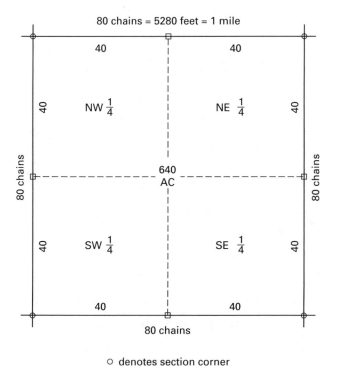

○ denotes section corner
□ denotes quarter corner

***Figure 13.1** Regular Section*

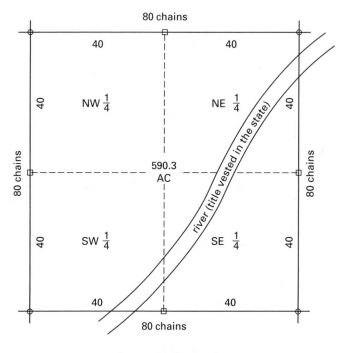

○ denotes section corner
□ denotes quarter corner

***Figure 13.2** Fractional Section*

3. Establishment of Quadrangles

Public land surveys of an area began with the establishment of an *initial point* (IP), the *meridian*, and the *baseline*. The IP location was selected for its convenience and adaptability for surveying away from it in the four cardinal directions. From this point, the meridian was established by running true (astronomic) north and south. This line is called the *principal meridian*, and monuments were set at 40-chain (1/2 mi) intervals to establish it. Such monuments are the regular quarter, section, and township corners. Townships are 6 mi on a side and contain 36 sections.

Sections are described by section number, township, range, and meridian. Townships are numbered north and south from the baseline, and ranges are numbered east and west from principal meridian (Figs. 13.3, 13.4, and 13.5).

The baseline is run true east and west from the initial point on a line of latitude (curved line). Monuments are set at 40-chain intervals and are regular quarter, section, and township corners. At every 24-mile interval along the principal meridian, another east-west line is established. These are called *standard parallels* or *correction lines* and are also called *curved lines*. At every 24-mile interval along the baseline or standard parallel, lines are run on a true north bearing and are called *guide meridians*. When the standard parallels and guide meridians are established, every quarter, section, and township corner is set on the ground. When a guide meridian intersects with a baseline or standard parallel, a closing corner is established by intersection and is tied by distance to the nearest corner on the

***Figure 13.4** Relationship of the Three Meridians Controlling Public Land Surveys in California*

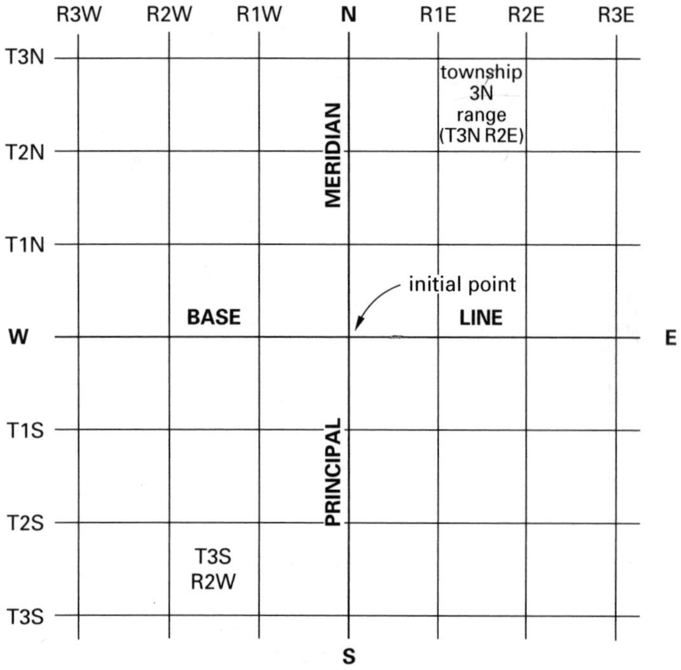

***Figure 13.3** Relationship of Township Numbering to the Baseline and Meridian*

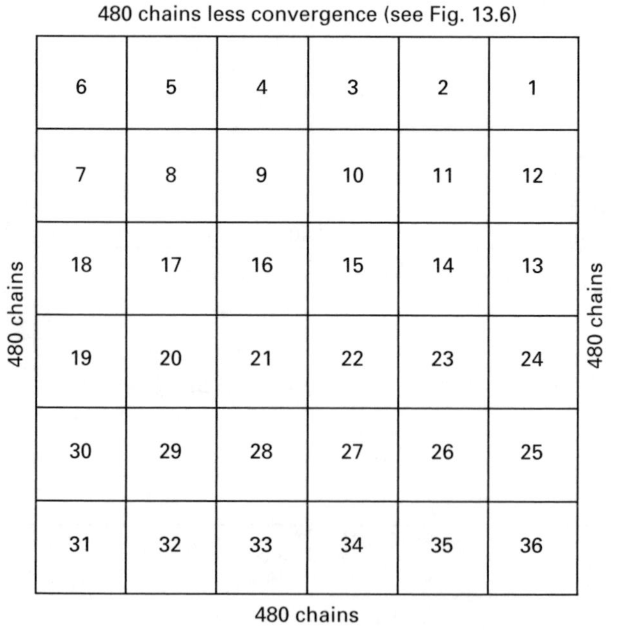

***Figure 13.5** Numbering of Sections in a Regular Township*

line closed upon. The monuments initially placed on the parallels or baseline control surveys to the north only and are called *standard corners*. Because meridians converge (these are lines of longitude), the distance of 24 miles between meridian lines on the baseline will be considerably less on the next parallel 24 miles north. When the guide meridians and baselines or parallels have been completed, 24-mile square subdivisions known as *quadrangles* containing 16 townships will exist on the ground.

All the controlling monuments, such as quarter, section, and township corners, will be in place, with the exception of the corners along the north line of the townships bordering a baseline or standard parallel. The standard corners set along these lines control the surveys north of these line and the northern limit of the quadrangles to the south (Fig. 13.6).

Practice Problems

1. A regular section in the USPLSS contains which of the following?

(A) 36 townships

(B) 160 acres

(C) 640 acres

(D) 7 lots

2. Which of the following is the standard unit of measurement in the USPLSS?

(A) link

(B) Gunter's chain

(C) U.S. standard foot

(D) meter

3. Which of the following is contained by a parcel of land that is 14.3 chains by 23.6 chains?

(A) 337.48 acres

(B) 33.75 acres

(C) 33.75 square chains

(D) 1,600,000 square feet

4. Which of the following is measured by the south line of a township along a standard parallel?

(A) 6 miles

(B) 480 chains

(C) 31,680.00 feet

(D) all of the above

5. Why might the north line of a township measure less than the south line?

(A) poor surveying techniques

(B) convergence of lines of longitude

(C) measurement on a curved surface

(D) all of the above

6. Why are standard parallels or correction lines established?

(A) so errors in the survey can be discovered

(B) because the law provides that only 16 townships be created in a quadrangle

(C) so township dimensions do not get too small

(D) to allow the surveyor to rest before continuing the survey

Problems 7 through 12 refer to the following figure.

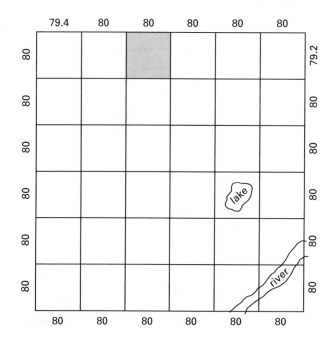

7. What number is the shaded section?

(A) Section 3

(B) Section 1E

(C) Section 1W

(D) Section 4

8. How many fractional sections are in this township?

(A) 11

(B) 12

(C) 15

(D) 16

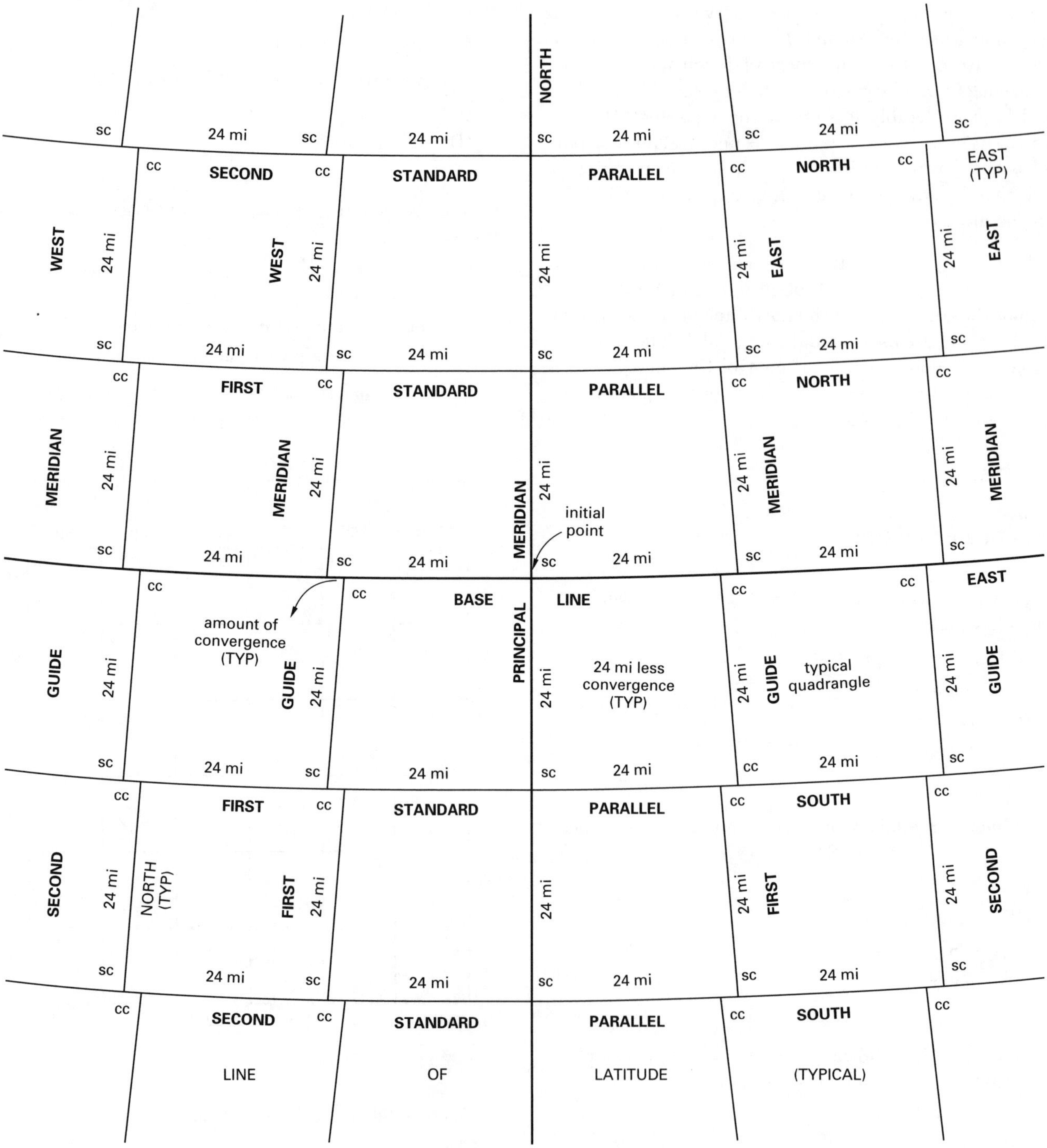

cc denotes closing corner

sc denotes standard corner

Figure 13.6 *Relationship of Quadrangles to the Meridian and Baseline*

9. What is the length of the north line of Lot 4, Section 6?

    (A) 19.4 chains

    (B) 19.85 chains

    (C) 20 chains

    (D) 40 chains

10. What is the south line of the township?

    (A) the 17th standard parallel north

    (B) the 4th standard parallel north

    (C) a correction line

    (D) a line of longitude

11. The west line of the township is which of the following?

    (A) 156 miles east of the principal meridian

    (B) 12 miles west of the 7th guide meridian east

    (C) a line of longitude

    (D) all of the above

12. Which of the following are the section corners along the south line of the township?

    (A) closing corners

    (B) correction corners

    (C) protracted corners

    (D) standard corners

Solutions

1. Answer (C)

2. Answer (B)

3.
$$\frac{(14.3 \text{ chains})(23.6 \text{ chains})}{10 \frac{\text{sq. chains}}{\text{ac}}} = 33.75 \text{ ac}$$

Answer (B)

4. Answer (D)

5. The east and west lines of a township run true north; therefore they are lines of longitude.

Answer (B)

6. The further north from the baseline, the smaller the townships measure because of convergence.

Answer (C)

7. Use Figure 13.5 to get Section 4.

Answer (D)

8. The township includes sections 1 through 6, 7, 18, 19, 30, 31, 23, 25, 35, and 36.

Answer (C)

9. Excess or deficiency is placed in the west 20 chains of the township when it is subdivided by protraction.

Answer (A)

10. Answer (B)

11. Answer (D)

12. Section corners are set on a standard parallel.

Answer (D)

Appendix: Stadia

Stadia Reductions

minutes	0° horizontal	0° vertical	1° horizontal	1° vertical	2° horizontal	2° vertical	3° horizontal	3° vertical
	distance							
0	100.00	0.00	99.97	1.74	99.88	3.49	99.73	5.23
2	100.00	0.06	99.97	1.80	99.87	3.55	99.72	5.28
4	100.00	0.12	99.97	1.86	99.87	3.60	99.71	5.34
6	100.00	0.17	99.96	1.92	99.87	3.66	99.71	5.40
8	100.00	0.23	99.96	1.98	99.86	3.72	99.70	5.46
10	100.00	0.29	99.96	2.04	99.86	3.78	99.69	5.52
12	100.00	0.35	99.96	2.09	99.85	3.84	99.69	5.57
14	100.00	0.41	99.95	2.15	99.85	3.89	99.68	5.63
16	100.00	0.47	99.95	2.21	99.84	3.95	99.68	5.69
18	100.00	0.52	99.95	2.27	99.84	4.01	99.67	5.75
20	100.00	0.58	99.95	2.33	99.83	4.07	99.66	5.80
22	100.00	0.64	99.94	2.38	99.83	4.13	99.66	5.86
24	100.00	0.70	99.94	2.44	99.82	4.18	99.65	5.92
26	99.99	0.76	99.94	2.50	99.82	4.24	99.64	5.98
28	99.99	0.81	99.93	2.56	99.81	4.30	99.63	6.04
30	99.99	0.87	99.93	2.62	99.81	4.36	99.63	6.09
32	99.99	0.93	99.93	2.67	99.80	4.42	99.62	6.15
34	99.99	0.99	99.93	2.73	99.80	4.47	99.61	6.21
36	99.99	1.05	99.92	2.79	99.79	4.53	99.61	6.27
38	99.99	1.11	99.92	2.85	99.79	4.59	99.60	6.32
40	99.99	1.16	99.92	2.91	99.78	4.65	99.59	6.38
42	99.99	1.22	99.91	2.97	99.78	4.71	99.58	6.44
44	99.98	1.28	99.91	3.02	99.77	4.76	99.58	6.50
46	99.98	1.34	99.90	3.08	99.77	4.82	99.57	6.56
48	99.98	1.40	99.90	3.14	99.76	4.88	99.56	6.61
50	99.98	1.45	99.90	3.20	99.76	4.94	99.55	6.67
52	99.98	1.51	99.89	3.26	99.75	4.99	99.55	6.73
54	99.98	1.57	99.89	3.31	99.74	5.05	99.54	6.79
56	99.97	1.63	99.89	3.37	99.74	5.11	99.53	6.84
58	99.97	1.69	99.88	3.43	99.73	5.17	99.52	6.90
60	99.97	1.74	99.88	3.49	99.73	5.23	99.51	6.96
$C = 0.75$	0.75	0.01	0.75	0.02	0.75	0.03	0.75	0.05
$C = 1.00$	1.00	0.01	1.00	0.03	1.00	0.04	1.00	0.06
$C = 2.15$	1.25	0.02	1.25	0.03	1.25	0.05	1.25	0.08

Stadia Reductions (continued)

minutes	4° horizontal	4° vertical	5° horizontal	5° vertical	6° horizontal	6° vertical	7° horizontal	7° vertical
				distance				
0	99.51	6.96	99.24	8.68	98.91	10.40	98.51	12.10
2	99.51	7.02	99.23	8.74	98.90	10.45	98.50	12.15
4	99.50	7.07	99.22	8.80	98.88	10.51	98.49	12.21
6	99.49	7.13	99.21	8.85	98.87	10.57	98.47	12.27
8	99.48	7.19	99.20	8.91	98.86	10.62	98.46	12.32
10	99.47	7.25	99.19	8.97	98.85	10.68	98.44	12.38
12	99.46	7.30	99.18	9.03	98.83	10.74	98.43	12.43
14	99.46	7.36	99.17	9.08	98.82	10.79	98.41	12.49
16	99.45	7.42	99.16	9.14	98.81	10.85	98.40	12.55
18	99.44	7.48	99.15	9.20	98.80	10.91	98.39	12.60
20	99.43	7.53	99.14	9.25	98.78	10.96	98.37	12.66
22	99.42	7.59	99.13	9.31	98.77	11.02	98.36	12.72
24	99.41	7.65	99.11	9.37	98.76	11.08	98.34	12.77
26	99.40	7.71	99.10	9.43	98.74	11.13	98.33	12.83
28	99.39	7.76	99.09	9.48	98.73	11.19	98.31	12.88
30	99.38	7.82	99.08	9.54	98.72	11.25	98.30	12.94
32	99.38	7.88	99.07	9.60	98.71	11.30	98.28	13.00
34	99.37	7.94	99.06	9.65	98.69	11.36	98.27	13.05
36	99.36	7.99	99.05	9.71	98.68	11.42	98.25	13.11
38	99.35	8.05	99.04	9.77	98.67	11.47	98.24	13.17
40	99.34	8.11	99.03	9.83	98.65	11.53	98.22	13.22
42	99.33	8.17	99.01	9.88	98.64	11.59	98.20	13.28
44	99.32	8.22	99.00	9.94	98.63	11.64	98.19	13.33
46	99.31	8.28	98.99	10.00	98.61	11.70	98.17	13.39
48	99.30	8.34	98.98	10.05	98.60	11.76	98.16	13.45
50	99.29	8.40	98.97	10.11	98.58	11.81	98.14	13.50
52	99.28	8.45	98.96	10.17	98.57	11.87	98.13	13.56
54	99.27	8.51	98.94	10.22	98.56	11.93	98.11	13.61
56	99.26	8.57	98.93	10.28	98.54	11.98	98.10	13.67
58	99.25	8.63	98.92	10.34	98.53	12.04	98.08	13.73
60	99.24	8.68	98.91	10.40	98.51	12.10	98.06	13.78
$C = 0.75$	0.75	0.06	0.75	0.07	0.75	0.08	0.74	0.10
$C = 1.00$	1.00	0.08	1.00	0.10	0.99	0.11	0.99	0.13
$C = 1.25$	1.25	0.10	1.24	0.12	1.24	0.14	1.24	0.16

Stadia Reductions (continued)

minutes	8° horizontal	8° vertical	9° horizontal	9° vertical	10° horizontal	10° vertical	11° horizontal	11° vertical
0	98.06	13.78	97.55	15.45	96.98	17.10	96.36	18.73
2	98.05	13.84	97.53	15.51	96.96	17.16	96.34	18.78
4	98.03	13.89	97.52	15.56	96.94	17.21	96.32	18.84
6	98.01	13.95	97.50	15.62	96.92	17.26	96.29	18.89
8	98.00	14.01	97.48	15.67	96.90	17.32	96.27	18.95
10	97.98	14.06	97.46	15.73	96.88	17.37	96.25	19.00
12	97.97	14.12	97.44	15.78	96.86	17.43	96.23	19.05
14	97.95	14.17	97.43	15.84	96.84	17.48	96.21	19.11
16	97.93	14.23	97.41	15.89	96.82	17.54	96.18	19.16
18	97.92	14.28	97.39	15.95	96.80	17.59	96.16	19.21
20	97.90	14.34	97.37	16.00	96.78	17.65	96.14	19.27
22	97.88	14.40	97.35	16.06	96.76	17.70	96.12	19.32
24	97.87	14.45	97.33	16.11	96.74	17.76	96.09	19.38
26	97.85	14.51	97.31	16.17	96.72	17.81	96.07	19.43
28	97.83	14.56	97.29	16.22	96.70	17.86	96.05	19.48
30	97.82	14.62	97.28	16.28	96.68	17.92	96.03	19.54
32	97.80	14.67	97.26	16.33	96.66	17.97	96.00	19.59
34	97.78	14.73	97.24	16.39	96.64	18.03	95.98	19.64
36	97.76	14.79	97.22	16.44	96.62	18.08	95.96	19.70
38	97.75	14.84	97.20	16.50	96.60	18.14	95.73	19.75
40	97.73	14.90	97.18	16.55	96.57	18.19	95.91	19.80
42	97.71	14.95	97.16	16.61	96.55	18.24	95.89	19.86
44	97.69	15.01	97.14	16.66	96.53	18.30	95.86	19.91
46	97.68	15.06	97.12	16.72	96.51	18.35	95.84	19.96
48	97.66	15.12	97.10	16.77	96.49	14.41	95.82	20.02
50	97.64	15.17	97.08	16.83	96.47	18.46	95.79	20.07
52	97.62	15.23	97.06	16.88	96.45	18.51	95.77	20.12
54	97.61	15.28	97.04	16.94	96.42	18.57	95.75	20.18
56	97.59	15.34	97.02	16.99	96.40	18.62	95.72	20.23
58	97.57	15.40	97.00	17.05	96.38	18.68	95.70	20.28
60	97.55	15.45	96.98	17.10	96.36	18.73	95.68	20.34
$C = 0.75$	0.74	0.11	0.74	0.12	0.74	0.14	0.73	0.15
$C = 1.00$	0.99	0.15	0.99	0.17	0.98	0.18	0.98	0.20
$C = 1.25$	1.24	0.18	1.23	0.21	1.23	0.23	1.22	0.25

Stadia Reductions (continued)

minutes	12° horizontal	vertical	13° horizontal	vertical	14° horizontal	vertical	15° horizontal	vertical
					distance			
0	95.68	20.34	94.94	21.92	94.15	23.47	93.30	25.00
2	95.65	20.39	94.91	21.97	94.12	23.52	93.27	25.05
4	95.63	20.44	94.89	22.02	94.09	23.58	93.24	25.10
6	95.61	20.50	94.86	22.08	94.07	23.63	93.21	25.15
8	95.58	20.55	94.84	22.13	94.04	23.68	93.18	25.20
10	95.56	20.60	94.81	22.18	94.01	23.73	93.16	25.25
12	95.53	20.66	94.79	22.23	93.98	23.78	93.13	25.30
14	94.51	20.71	94.76	22.28	93.95	23.83	93.10	25.35
16	95.49	20.76	94.73	22.34	93.93	23.88	93.07	25.40
18	95.46	20.81	94.71	22.39	93.90	23.93	93.04	25.45
20	95.44	20.87	94.68	22.44	93.87	23.99	93.01	25.50
22	95.41	20.92	94.66	22.49	93.84	24.04	92.98	25.55
24	95.39	20.97	94.63	22.54	93.82	24.09	92.95	25.60
26	95.36	21.03	64.60	22.60	93.79	24.14	92.92	25.65
28	95.34	21.08	94.58	22.65	93.76	24.19	92.89	25.70
30	95.32	21.13	94.55	22.70	93.73	24.24	92.86	25.75
32	95.29	21.18	94.52	22.75	93.70	24.29	92.83	25.80
34	95.27	21.24	94.50	22.80	93.67	24.34	92.80	25.85
36	95.24	21.29	94.47	22.85	93.65	24.39	92.77	25.90
38	95.22	21.34	94.44	22.91	93.62	24.44	92.74	25.95
40	95.19	21.39	94.42	22.96	93.59	24.49	92.71	26.00
42	95.17	21.45	94.39	23.01	93.56	24.55	92.68	26.05
44	95.14	21.50	94.36	23.06	93.53	24.60	92.65	26.10
46	95.12	21.55	94.24	23.11	93.50	24.65	92.62	26.15
48	95.09	21.60	94.31	23.16	93.47	24.70	92.59	26.20
50	95.07	21.66	94.28	23.22	93.45	24.75	92.56	26.25
52	95.04	21.71	94.26	23.27	93.42	24.80	92.53	26.30
54	95.02	21.76	94.23	23.32	93.39	24.85	92.49	26.35
56	94.99	21.81	94.20	23.37	93.36	24.90	92.46	26.40
58	94.97	21.87	94.17	23.42	93.33	24.95	92.43	26.45
60	94.94	21.92	94.15	23.47	93.30	25.00	92.40	26.50
$C = 0.75$	0.73	0.16	0.73	0.18	0.73	0.19	0.72	0.20
$C = 1.00$	0.98	0.22	0.97	0.23	0.97	0.25	0.96	0.27
$C = 1.25$	1.22	0.27	1.22	0.29	1.21	0.31	1.20	0.33

Stadia Reductions (continued)

minutes	16° horizontal	16° vertical	17° horizontal	17° vertical	18° horizontal	18° vertical	19° horizontal	19° vertical
	distance							
0	92.40	26.50	91.45	27.96	90.45	29.39	89.40	30.78
2	92.37	26.55	91.42	28.01	90.42	29.44	89.36	30.83
4	92.34	26.59	91.39	28.06	90.38	29.48	89.33	30.87
6	92.31	26.64	91.35	28.10	90.35	29.53	89.29	30.92
8	92.28	26.69	91.32	28.15	90.31	29.58	89.26	30.97
10	92.25	26.74	91.29	28.20	90.28	29.62	89.22	31.01
12	92.22	26.79	91.26	28.25	90.24	29.67	89.18	31.06
14	92.19	26.84	91.22	28.30	90.21	29.72	89.15	31.10
16	92.15	26.89	91.19	28.34	90.18	29.76	89.11	31.15
18	92.12	26.94	91.16	28.39	90.14	29.81	89.08	31.19
20	92.09	26.99	91.12	28.44	90.11	29.86	89.04	31.24
22	92.06	27.04	91.09	28.49	90.07	29.90	89.00	31.28
24	92.03	27.09	91.06	28.54	90.04	29.95	88.97	31.33
26	92.00	27.13	91.02	28.58	90.00	30.00	88.93	31.38
28	91.97	27.18	90.99	28.63	89.97	30.04	88.89	31.42
30	91.93	27.23	90.96	28.68	89.93	30.09	88.86	31.47
32	91.90	27.28	90.92	28.73	89.90	30.14	88.82	31.51
34	91.87	27.33	90.89	28.77	89.86	30.18	88.78	31.56
36	91.84	27.38	90.86	28.82	89.83	30.23	88.75	31.60
38	81.81	27.43	90.82	28.87	89.79	30.28	88.71	31.65
40	91.77	27.48	90.79	28.92	89.76	30.32	88.67	31.69
42	91.74	27.52	90.76	28.96	89.72	30.37	88.64	31.74
44	91.71	27.57	90.72	29.01	89.69	30.41	88.60	31.78
46	91.68	27.62	90.69	29.06	89.65	30.46	88.56	31.83
48	91.65	27.67	90.66	29.11	89.61	30.51	88.53	31.87
50	91.61	27.72	90.62	29.15	89.58	30.55	88.49	31.92
52	91.58	27.77	90.59	29.20	89.54	30.60	88.45	31.96
54	91.55	27.81	90.55	29.25	89.51	30.65	88.41	32.01
56	91.52	27.86	90.52	29.30	89.47	30.69	88.38	32.05
58	91.48	27.91	90.49	29.34	89.44	30.74	88.34	32.09
60	91.45	27.96	90.45	29.39	89.40	30.78	88.30	32.14
$C = 0.75$	0.72	0.21	0.72	0.23	0.71	0.24	0.71	0.25
$C = 1.00$	0.96	0.28	0.95	0.30	0.95	0.32	0.94	0.33
$C = 1.25$	1.20	0.36	1.19	0.38	1.19	0.40	1.18	0.42

Stadia Reductions *(continued)*

minutes	20° horizontal	20° vertical	21° horizontal	21° vertical	22° horizontal	22° vertical	23° horizontal	23° vertical
				distance				
0	88.30	32.14	87.16	33.46	85.97	34.73	84.73	35.97
2	88.26	32.18	87.12	33.50	85.93	34.77	84.69	36.01
4	88.23	32.23	87.08	33.54	85.89	34.82	84.65	36.05
6	88.19	32.27	87.04	33.59	85.85	34.86	84.61	36.09
8	88.15	32.32	87.00	33.63	85.80	34.90	84.57	36.13
10	88.11	32.36	86.96	33.67	85.76	34.94	84.52	36.17
12	88.08	32.41	86.92	33.72	85.72	34.98	84.48	36.21
14	88.04	32.45	86.88	33.76	85.68	35.02	84.44	36.25
16	88.00	32.49	86.84	33.80	85.64	35.07	84.40	36.29
18	87.96	32.54	86.80	33.84	85.60	35.11	84.35	36.33
20	87.93	32.58	86.77	33.89	85.56	35.15	84.31	36.37
22	87.89	32.63	86.73	33.93	85.52	35.19	84.27	36.41
24	87.85	32.67	86.69	33.97	85.48	35.23	84.23	36.45
26	87.81	32.72	86.65	34.01	85.44	35.27	84.18	36.49
28	87.77	32.76	86.61	34.06	85.40	35.31	84.14	36.53
30	87.74	32.80	86.57	34.10	85.36	35.36	84.10	36.57
32	87.70	32.85	86.53	34.14	85.31	35.40	84.06	36.61
34	87.66	32.89	86.49	34.18	85.27	35.44	84.01	36.65
36	87.62	32.93	86.45	34.23	85.23	35.48	83.97	36.69
38	87.58	32.98	86.41	34.27	85.19	35.52	83.93	36.73
40	87.54	33.02	86.37	34.31	85.15	35.56	83.89	36.77
42	87.51	33.07	86.33	34.35	85.11	35.60	83.84	36.80
44	87.47	33.11	86.29	34.40	85.07	35.64	83.80	36.84
46	87.43	33.15	86.25	34.44	85.02	35.68	83.76	36.88
48	87.39	33.20	86.21	34.48	84.98	35.72	83.72	36.92
50	87.35	33.24	86.17	34.52	84.94	35.76	83.67	36.96
52	87.31	33.28	86.13	34.57	84.90	35.80	83.63	37.00
54	87.27	33.33	86.09	34.61	84.86	35.85	83.59	37.04
56	87.24	33.37	86.05	34.65	84.82	35.89	83.54	37.08
58	87.20	33.41	86.01	34.69	84.77	35.93	83.50	37.12
60	87.16	33.46	85.97	34.73	84.73	35.97	83.46	37.16
$C = 0.75$	0.70	0.26	0.70	0.27	0.69	0.29	0.69	0.30
$C = 1.00$	0.94	0.35	0.93	0.37	0.92	0.38	0.92	0.40
$C = 1.25$	1.17	0.44	1.16	0.46	1.15	0.48	1.15	0.50

Stadia Reductions (continued)

minutes	24° horizontal	24° vertical	25° horizontal	25° vertical	26° horizontal	26° vertical	27° horizontal	27° vertical
				distance				
0	83.46	37.16	82.14	38.30	80.78	39.40	79.39	40.45
2	83.41	37.20	82.09	38.34	80.74	39.44	79.34	40.49
4	83.37	37.23	82.05	38.38	80.69	39.47	79.30	40.52
6	83.33	37.27	82.01	38.41	80.65	39.51	79.25	40.55
8	83.28	37.31	81.96	38.45	80.60	39.54	79.20	40.59
10	83.24	37.35	81.92	38.49	80.55	39.58	79.15	40.62
12	83.20	37.39	81.87	38.53	80.51	39.61	79.11	40.66
14	83.15	37.43	81.83	38.56	80.46	39.65	79.06	40.69
16	83.11	37.47	81.78	38.60	80.41	39.69	79.01	40.72
18	83.07	37.51	81.74	38.64	80.37	39.72	78.96	40.76
20	83.02	37.54	81.69	38.67	80.32	39.76	78.92	40.79
22	82.98	37.58	81.65	38.71	80.28	39.79	78.87	40.82
24	82.93	37.62	81.60	38.75	80.23	39.83	78.82	40.86
26	82.89	37.66	81.56	38.78	80.18	39.86	78.77	40.89
28	82.85	37.70	81.51	38.82	80.14	39.90	78.73	40.92
30	82.80	37.74	81.47	38.86	80.09	39.93	78.68	40.96
32	82.76	37.77	81.42	38.89	80.04	39.97	78.63	40.99
34	82.72	37.81	81.38	38.93	80.00	40.00	78.58	41.02
36	82.67	37.85	81.33	38.97	79.95	40.04	78.54	41.06
38	82.63	37.89	81.28	39.00	79.90	40.07	78.49	41.09
40	82.58	37.93	81.24	39.04	79.86	40.11	78.44	41.12
42	82.54	37.96	81.19	39.08	79.81	40.14	78.39	41.16
44	82.49	38.00	81.15	39.11	79.76	40.18	78.34	41.19
46	82.45	38.04	81.10	39.15	79.72	40.21	78.30	41.22
48	82.41	38.08	81.06	39.18	79.67	40.24	78.25	41.26
50	82.36	38.11	81.01	39.22	79.62	40.28	78.20	41.29
52	82.32	38.15	80.97	39.26	79.58	40.31	78.15	41.32
54	82.27	38.19	80.92	39.29	79.53	40.35	78.10	41.35
56	82.23	38.23	80.87	39.33	79.48	40.38	78.06	41.39
58	82.18	38.26	80.83	39.36	79.44	40.42	78.01	41.42
60	82.14	38.30	80.78	39.40	79.39	40.45	77.96	41.45
$C = 0.75$	0.68	0.31	0.68	0.32	0.67	0.33	0.67	0.35
$C = 1.00$	0.91	0.41	0.90	0.43	0.89	0.45	0.89	0.46
$C = 1.25$	1.14	0.52	1.13	0.54	1.12	0.56	1.11	0.58

Professional Publications, Inc. • Belmont, California

Stadia Reductions (continued)

minutes	28°		29°		30°	
	distance					
	horizontal	vertical	horizontal	vertical	horizontal	vertical
0	77.96	41.45	76.50	42.40	75.00	43.30
2	77.91	41.48	76.45	42.43	74.95	43.33
4	77.86	41.52	76.40	42.46	74.90	43.36
6	77.81	41.55	76.35	42.49	74.85	43.39
8	77.77	41.58	76.30	42.53	74.80	43.42
10	77.72	41.61	76.25	42.56	74.75	43.45
12	77.67	41.65	76.20	42.59	74.70	43.47
14	77.62	41.68	76.15	42.62	74.65	43.50
16	77.57	41.71	76.10	42.65	74.60	43.53
18	77.52	41.74	76.05	42.68	74.55	43.56
20	77.48	41.77	76.00	42.71	74.49	43.59
22	77.42	41.81	75.95	42.74	74.44	43.62
24	77.38	41.84	75.90	42.77	74.39	43.65
26	77.33	41.87	75.85	42.80	74.34	43.67
28	77.28	41.90	75.80	42.83	74.29	43.70
30	77.23	41.93	75.75	42.86	74.24	43.73
32	77.18	41.97	75.70	42.89	74.19	43.76
34	77.13	42.00	75.65	42.92	74.14	43.79
36	77.09	42.03	75.60	42.95	74.09	43.82
38	77.04	42.06	75.55	42.98	74.04	43.84
40	76.99	42.09	75.50	43.01	73.99	43.87
42	76.94	42.12	75.45	43.04	73.93	43.90
44	76.89	42.15	75.40	43.07	73.88	43.93
46	76.84	42.19	75.35	43.10	73.83	43.95
48	76.79	42.22	75.30	43.13	73.78	43.98
50	76.74	42.25	75.25	43.16	73.73	44.01
52	76.69	42.28	75.20	43.18	73.68	44.04
54	76.64	42.31	75.15	43.21	73.63	44.07
56	76.59	42.34	75.10	43.24	73.58	44.09
58	76.55	42.37	75.05	43.27	73.52	44.12
60	76.50	42.40	75.00	43.30	73.47	44.15
$C = 0.75$	0.66	0.36	0.65	0.37	0.65	0.38
$C = 1.00$	0.88	0.48	0.87	0.49	0.86	0.51
$C = 1.25$	1.10	0.60	1.09	0.62	1.08	0.63

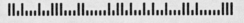